**Field Measurement Methods
in Soil Science**

S. Wessel-Bothe and L. Weihermüller (Eds.)

Field Measurement Methods in Soil Science

edited by

S. Wessel-Bothe and L. Weihermüller

With 76 figures and 18 tables

Borntraeger Science Publishers 2020

Wessel-Bothe, S. and Weihermüller, L. (Eds.): Field Measurement Methods in Soil Science

Editors:
Stefan Wessel-Bothe, ecoTech Umwelt-Meßsysteme GmbH, Klara-M.-Faßbinder-Straße 1A, 53121 Bonn, Germany. soil@ecoTech-Bonn.de
Lutz Weihermüller, Forschungszentrum Jülich, IBG-3, Wilhelm-Johnen-Straße, 52428 Jülich, Germany.
l.weihermueller@fz-juelich.de

We would be pleased to receive your comments on the content of this book:
editors@schweizerbart.de

Cover: Photos by Stefan Wessel-Bothe.

This publication has been made possible with the generous support by

ISBN 978-3-443-01109-3

Information on this title: www.borntraeger-cramer.de/9783443011093
© 2020 Gebr. Borntraeger Verlagsbuchhandlung, Stuttgart, Germany

All rights reserved. No part of this publication may be reproduced, stored in a retrieval system, or transmitted, in any form or by any means, electronic, mechanical photocopying, recording, or otherwise, without the prior written permission of Gebr. Borntraeger Science Publishers.

Publisher: Gebr. Borntraeger Verlagsbuchhandlung
             Johannesstraße 3A, 70176 Stuttgart, Germany
             mail@borntraeger-cramer.de   www.borntraeger-cramer.de

⊚ Printed on permanent paper conforming to ISO 9706-1994
Printed in Germany by Gulde Druck, Tübingen

Preface

Soils are progressively recognized as critical environmental compartment due to the multitude of functions they provide as habitat, as medium for the production of food, fibre and bio-energy, as regulators of the cycling of energy, water and elements such as carbon and nitrogen in ecosystems, as filters and buffers for pollutants, and their role as archives of human and landscape history. Consequently, an increasingly broad spectrum of practitioners and scientists addresses soil issues and works with soils in the context of their specific field of interest and profession. The rising interest in soils drives a mounting demand for soil information, which frequently has to be generated using field measurements. When planning these field measurements, many questions arise: What is the particular question that has to be answered? Which specific parameters do I have to measure for answering this question? Which method and which equipment is best suited to record these parameters at which precision and cost? Which preconditions have to be met for using a certain method or equipment in a meaningful way? How is the equipment installed and used? How many replicates are required? Which model cases exist for the application of certain methods? Which pitfalls lurk when using a certain method? How are the generated data interpreted and quality checked?

Much know-how to answer these questions exists distributed for example in research institutes and extension services. However, the number of people in these institutions and their available time for supporting the planning of soil-related field measurements is increasingly limited. Therefore, the book at hand fills a critical gap by providing hands-on support for the design and operation of field measurements in soil science. The book is useful not only for soil science-"beginners" searching for an introductory overview of available techniques, but also for more experienced colleagues by providing "best practice" guidelines for comparable installation and operation of field instruments. Most likely, the book cannot substitute personal discussions and consultation, but it can make these discussions much more efficient and productive.

I thank all readers for their commitment to our precious soil resource. May this book contribute to the successful generation of soil information needed for the preservation of intact and functional soils for future generations.

Giessen, July 2019 *Prof. Dr. Jan Siemens*

The Authors

Prof. Dr. **Tim Mansfeldt** is head of the Soil Geography/Soil Science group at the University of Cologne, Faculty of Mathematics and Natural Sciences, Department of Geosciences. His research focuses, among others, on the identification and characterization of reducing conditions in soils.

Prof. Dr. **Sören Thiele-Bruhn** is head of the Soil Science department, Regional and Environmental Sciences, Soil Science, University of Trier. He especially works in soil ecology and soil toxicology on the fate and effects of natural and synthetic compounds and substrates in interaction with soil microbiota.

PD Dr. **Lutz Weihermüller** is a senior scientist at the Agrosphere Institute IBG-3 at the Forschungszentrum Jülich GmbH and responsible for the soil physical laboratories. He specialized in numerical modelling for various applications and performed a wide range of field experiments.

PD Dr. **Heye Bogena** is a senior scientist at the Agrosphere Institute IBG-3 at the Forschungszentrum Jülich GmbH. He leads the research group "Terrestrial observation platforms" and coordinates the Helmholtz project TERENO (Terrestrial Environmental Observatories). His research includes wireless sensor networks, soil moisture sensing, and a wide range of topics in soil and catchment hydrology.

Dr. **Axel Lamparter** is working at the Federal Institute for Geosciences and Natural Resources (BGR) – Sub-Department 2.4 – Soil as a Resource – Properties and Dynamics. He is responsible for the soil physics laboratory and skilled in performing field experiments and in-situ measurements. Additionally, he contributes to several projects at BGR.

Dr. **Raimund Schneider** is senior scientist at the Department of Soil Science, University of Trier, responsible for soil physical, mechanical, and micromorphological laboratories and specialized in soil protection, compaction, hydrology, and sustainable soil management in agriculture and forestry.

Jesús Rodrigo Comino, PhD, Physical Geography, Trier University, Soil Erosion and Degradation Research Group, Department of Geography, University of Valencia, 46010 Valencia, Spain.

Dr. **Dominic Demandis** a Postdoc at the Chair of Hydrology, University of Freiburg. He is specialized on soil hydrology with a focus on the spatial and temporal dynamics of soil water flow.

Dr. **Heinz Peter Schrey** as senior soil scientist is head of section Soil Information System, at the Geological Survey of Northrhine-Westfalia, specialized on web map services, soil protection, and quantifying codification for evaluation of soil maps.

Dr. **Stefan Wirtz** was a research associate at the Department of Physical Geography of Trier University and is now at the "Bundeswehr GeoInformation Service". His research field is the experimental investigation of rill erosion processes and the influence of different hydraulic parameters on soil erosion.

Dr. **Thomas Iserloh** is a research associate at the Department of Physical Geography of Trier University. As interdisciplinary-oriented environmental scientist with a PhD in Physical Geography, he has a strong background in the disciplines of Soil Science and Geomorphology. He focuses on field work, particularly on experimental soil erosion research.

Dr. **Miriam Marzen** is a research associate at the Department of Physical Geography of Trier University. Her research bases on geomorphological, soil- and biogeographic key topics within a socio-ecological context. She has a strong focus on empirical-experimental research concerning process understanding and quantification of soil erosion by wind and water in the context of land use change and land degradation.

Dr. **Wolfgang Fister** is a researcher and lecturer at the Department of Environmental Sciences at the University of Basel. He is currently working on various topics related to soil erosion in the "soil erosion and rainfall laboratory" in Basel and in the field. For example, the assessment of the efficacy of biochar to soil erosion, the measurement of raindrop impacted flow erosion, and the risk assessment of dust emissions from the Free State in South Africa are of special interest to him at present.

Dr. **Stefan Wessel-Bothe** is head of the Soil Science section at ecoTech Monitoring Systems GmbH in Bonn and specialized on water and solute transport processes in soil. He focuses on R&D to optimize and develop new scientific methods and is senior consultant for designing the experimental set-up of scientific projects.

Table of Contents

Preface (Jan Siemens) 5
The Authors 6

Chapter 1 **General introduction** (L. Weihermüller, S. Wessel-Bothe) 9

    1.1. Measurement location 9
    1.2. Number of stations 9
    1.3. Experiment design 10
    1.4. Sensor installation 13
    1.5. Arrangement of enclosures, cabinets, housings etc. 16
    1.6. Post-processing of aquired data 17
    1.7. Concluding remarks 17
    1.8. References 18

Soil physicochemical parameters

Chapter 2 **Soil redox potential** (T. Mansfeldt) 19

    2.1. Introduction 19
    2.2. Method selection 23
    2.3. Method application 25
    2.4. References 37

Chapter 3 **Soil pH-Value** (S. Thiele-Bruhn) 43

    3.1. Introduction 43
    3.2. Method selection 52
    3.3. Laboratory methods for measuring pH in soils 52
    3.4. Field methods of determining the soil pH 56
    3.5. References... 63

Soil water parameters

Chapter 4 **Soil water content measurement** (H. Bogena, L. Weihermüller) 69

    4.1. Introduction 69
    4.2. Soil water content sensor types – Method selection 69
    4.3. Implementation of soil water content sensors 75
    4.4. Related methods 77
    4.5. References 78

Chapter 5 **Soil matric potential/Soil water tension** (A. Lamparter) 83

    5.1. Introduction 83
    5.2. Selection of the appropriate method 86
    5.3. Implementation 95
    5.4. Related Methods 98
    5.5. References 98

Chapter 6 **In-situ soil water sampling** (L. Weihermüller) 99

    6.1. Introduction 99
    6.2. Suction cups 99
    6.3. Suction plates 108
    6.4. Pan lysimeters 113
    6.5. Capillary wicks 116
    6.6. Lysimeters 119
    6.7. Substance-specific requirements of solute sampling 124
    6.8. References 128

Chapter 7 **Infiltration and water conductivity in saturated and unsaturated soils**
(R. Schneider, J. Rodrigo Comino, D. Demand, H. Schrey) 133

    7.1. Introduction. 133
    7.2. Method selection. 133
    7.3. Method descriptions 134
    7.4. References. 157

Soil mechanical parameters

Chapter 8 **Experimental field methods to quantify soil erosion by water and wind-driven rain** (S. Wirtz, T. Iserloh, M. Marzen, W. Fister) 165

    8.1. Introduction 165
    8.2. Selecting the most appropriate method 167
    8.3. Experimental methods 169
    8.4. References 187

Chapter 9 **Penetration resistance** (H. Schrey) 191

    9.1. Introduction 191
    9.2. Selection of the most suitable method 193
    9.3. Using a penetrometer 198
    9.4. Data analysis 200
    9.5. References 206

Index 207

1. General Introduction

Lutz Weihermüller and Stefan Wessel-Bothe

Knowledge of the soil states (e.g., soil water content, soil pH, redox potential, matric potential) is imperative to reliably describe water and solute transport processes, changes in the soil compartment by e.g. human activities or climate change, to implement soil conservation measures, facilitate crop production, or for modeling purposes. Over the last decades, various new sensor types and measurement techniques were developed and sensor development is still an emerging field, resulting in improved or even new sensor techniques and types. Because of this, the user is more and more dependent on guidelines which sensor/technique is appropriate for the study planned.

As outlined in the documentation of the single sampling devices, sensors, and measurement techniques, the right choice of the measurement sensor/technique/sampling device is not an easy task. First of all, most sensors/techniques/sampling devices differ greatly with regard to the type of information they collect, their resolution in space and time, their cost and maintenance, as well as in the requirement of expert knowledge. Even if the individual chapters of this book try to point out advantages and disadvantages of each system/sensor, this should not be regarded as the *only recommendation* for later decisions. Moreover, this book should be conceived as a guideline for decision making taking all advantages/disadvantages, costs etc. into account, without losing the focus of the experiment planned.

1.1. Measurement location

In general, the scientist must ensure that the location of the soil monitoring station is well chosen, irrelevant if only one or several stations are to be distributed over an area or catena. To ensure that the selected location is representative for the study site, soil information should be gathered, which may be used for estimating the heterogeneity at the location. This information can be obtained from, for example, aerial views, topographical and fine scale soil maps, or information deduced from own mapping studies (e.g., Pürckhauer augering). Avoid previously disturbed areas, vicinity to buildings, and the influence of close by vegetation or surface waters which may have impact on the soil's hydrology. Influence factors such as these – if they are not part of the study – would interfere with your results in a misleading way.

1.2. Number of stations

In many cases, one is only able to operate one single monitoring station on one single site due to budget constraints. In these cases, it must be understood that i) the experimental design must be planned as meticulously as possible, and ii) that the results from one single site may only be representative of a limited area. The extent of transferability then depends on the landscape's particular features such as topography and the spatial heterogeneity of the soil.

If the budget allows setting up more than one site with the same instrumentation, the results so obtained will potentially be more representative of the area under investigation. Nevertheless, it is not always the best solution to have as many measuring points as possible within a study area. On the other hand, distributing only few sensors at various spots within the study area might cause a lack of information on the short term variability at each single location. Additionally, the reliability of sensor readings for each spot cannot be tested, unless replicate measurement data are available. A compromise would be to install fewer plots with a higher (vertical and horizontal) sensor density instead of installing too many plots within one single field with only sparse instrumentation. In some cases, the use of inexpensive but less sophisticated (less accurate) sensors may be an appropriate alternative. However, to obtain reliable results using low-cost sensors, it is always worth to consider the concessions concerning the quality of data from the sensors. In conclusion, lower quality readings are only acceptable when their statistic evaluation and the scientific goal will not be affected.

1.3. Experiment design

In a first step, the experimentalist should exactly define the study target and design the experiment appropriately *before* selecting any sampler or sensor. Based on the design and scientific questions, some sensor types or systems may be already identified. In the next step, different sensors providing the same information (e.g., water content information) should be reviewed and based on cost, maintenance, and existing expert knowledge an appropriate sensor/system should be chosen. In this context, maintenance denotes the sum of expenditure on human labor and service inspections, whereas the cost is the total of the maintenance, cost of acquisition, and installation of the sensors.

In conclusion, the decision on the appropriate sensor/system will be defined by i) the experimental target and ii) potential limitations in terms of installation effort, maintenance, financial background, and existing expert knowledge. In any case, the suitability of the sensor/ sampling system must be carefully deliberated.

1.3.1. Spatial heterogeneity

Because most sensors/techniques determine the soil state of interest (e.g., water content) in a relative small (measurement) volume, but most applications of soil research address the field scale, a mismatch of measurement scale and scale of interest poses a serious problem, because it is generally questionable whether point measurements are representative of the larger scale. From a scientific point of view, soil can be understood as a porous medium variable in space and time, whereby its state variables (e.g., soil water content, soil pH, redox potential, matric potential) are extensive quantities. When measuring such extensive quantities, one must acknowledge that the quantity depends on the measurement scale. To account for the scale effect Bear (1972) introduced the concept of the representative elementary volume (REV), which states that a volume is representative to a certain extent as long as the state does not fluctuate within this volume. Unfortunately, the complexity of this simple definition increases whenever it is applied to a real world problem. Firstly, the actual size of the REV depends on the system under investigation (this means that specific REV volumes may be applicable for each, simultaneously measured parameter), and secondly, the actual size of the REV depends on the areal size of the sampled region. As Durner and Flühler (2005) have stated, enlarging the averaging volume will lead to the inclusion of new, larger size structural elements. The dependency of a soil state (e.g., soil water content) on the characteristic length of the averaging volume is shown schematically in Fig. 1.1.

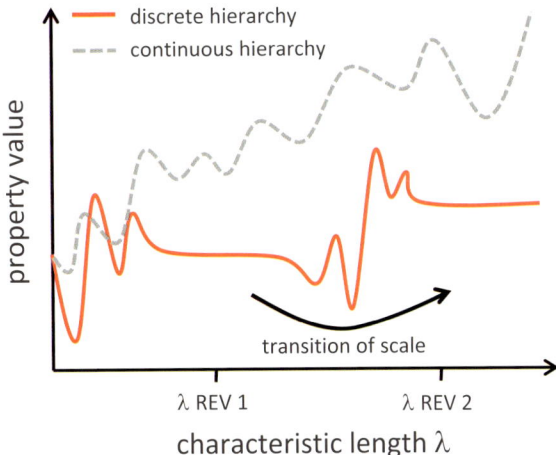

Fig. 1.1. Concept of the representative elementary volume (REV) assuming two different concepts of spatial heterogeneity (adopted from Zurmühl 1994). As can be seen, two different REVs are detectable (REV 1 and REV 2) for the discrete hierarchy at given characteristic length λ with two different property values. Moving from REV 1 to REV 2 changes the property value. On the other hand, a continuous hierarchy does not allow to delineate REVs.

Apparently, there is a clear disparity of hierarchical levels (scales), where repeated measurements (e.g., replicates in sensors) yield consistent values. In the transition zone of two hierarchical levels the measured value becomes unstable, which is indicated by increased variation of measured values (replicates).

1.3.2. Number of replicates

Because the natural variability of the soil states (often termed heterogeneity) is unknown, the experimentalist relies on a number of replicate measurements (sensors) per depth level. Unfortunately, the number of replications (n) is often falsely assumed to be sufficient if $n = 3$. As Weihermüller et al. (2005) nicely showed for suction cups sampling, n has to be far greater than 3 to estimate a mean solute breakthrough curve, even in a fairly homogeneous soil system. Therefore, the question arises how many replicates have to be installed? Logically, the required number can only be determined after analyzing ∞ replicates, which is neither feasible nor constructive. To get at least a hint of the number of replicates required, the literature should be screened for comparable setups and indications of heterogeneity. Because some parameters have a different distribution in soil than others (e.g., soil temperature has a generally lower vertical heterogeneity as soil water content), it is necessary to clarify the required number of replicates particular to the state of interest. It may be sufficient to install only one soil temperature sensor, while on the same site at least three sensors for water content are necessary. In most chapters of this book, general hints or a secret recommended number of replicates of instruments are presented for individual soil states of interest.

1.3.3. Precision, trueness and accuracy

The technical terms *precision*, *trueness*, and *accuracy* are often mixed up. In fact, the accuracy of a reading is a combination of its trueness and its precision. To clarify this, the definitions and differences of these terms are given below.

A) Precision

Precision or internal reproducibility is the degree of agreement between independent measurement results under given static conditions (e.g., constant soil water content or temperature). If the sensor output shows low variability in these conditions, the measurement method/sensor has high precision. On the other hand, it is important to note that a high precision does not mean that the measured values are correct. You could be precisely wrong (see Fig. 1.2).

B) Trueness

The trueness or external reproducibility denotes the degree agreement of the mean value from a large set of measured sensor data and the accepted reference value measured by a reference method. In other words, if the mean value of a large number of measurements is in good agreement with the reference value, the trueness is high. However, this procedure does not give any information about the scatter of the individual readings (see Fig. 1.2).

B) Accuracy

Accuracy is the degree of the correspondence between the (single) measurement result and the true value of the measured variable. A high accuracy can therefore only be achieved if both, the precision and the trueness, are high.

This relationship between precision, trueness, and accuracy can be visualized very well by a simple sketch of target symbols as shown in Fig. 1.2.

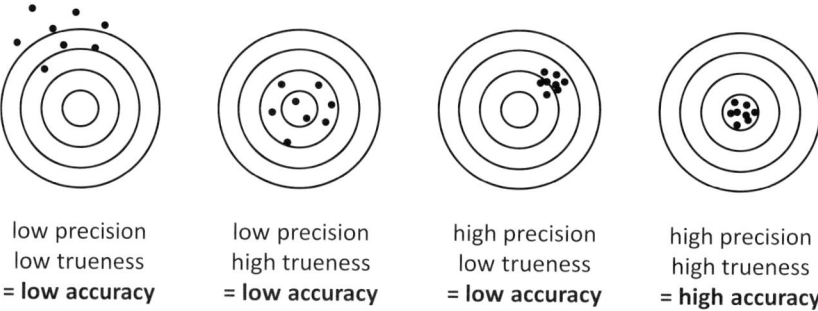

low precision	low precision	high precision	high precision
low trueness	high trueness	low trueness	high trueness
= low accuracy	= low accuracy	= low accuracy	= high accuracy

Fig. 1.2. Sensor/measurement accuracy as a function of precision and trueness. Note, that the center of the rings defines the real value.

Any measurement can only be accurate if it is precise *and* true. Looking at the given accuracy of a sensor/measuring system on one hand and the required accuracy for the study, the decision for high accuracy with high costs and low accuracy with low cost must be balanced carefully. In extreme cases, the instrumentation is unable to provide the information required and the success of the whole study might be affected by poor sensor instrumentation.

1.4. Sensor installation

Most of the chapters describing the sensors/devices/samplers provide links and procedures for the correct installation of the respective sensors. Nevertheless, some generally applicable remarks for the installation will be provided here.

1.4.1. Spatial resolution

1.4.1.1. Installation at a specific depth

Sensor or system installation at different depths often follows a classical scheme, where the sensors are equally distributed over a depth range (e.g., 30, 60, 90, 120 cm). Generally, most soil states unfortunately do not vary linearly with time and depth and most of variation in the soil states are often observed close to dynamic boundaries. These boundaries are the atmosphere with dynamic changes between precipitation and evapotranspiration but can also be dynamic water table fluctuations at greater depths. A sketch of the dynamic behavior with time is provided in Fig. 1.3 where soil state dynamics are depicted as a system response to atmospheric forcing (here atmospheric temperature changes).

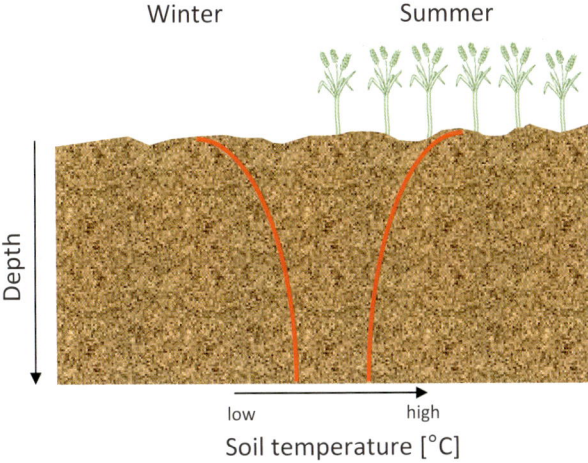

Fig. 1.3. Typical distribution of state values and dynamics over depth for two different situations. Here, soil temperature profile in winter and summer. The strongest gradients are observed at the soil/atmosphere interface and decrease with soil depth.

As can be exemplarily seen from the sketch in Fig. 1.3, the soil temperature distribution over depth varies extremely close to the surface between seasons and shows less difference at greater depth. Due to the direct control exerted on soil temperature by the atmosphere, other states behave similarly, e.g. soil water content/soil matric potential in a dry period and after a long rainfall period. Consequently, the top soil layer responds more dynamically and directly than do deeper zones and also shows the largest range of readings under such conditions. To monitor this general behavior properly, the sensor spacing should be denser close to the soil surface and at greater distances in deeper zones. Therefore, good experiences were made with non-uniform installation depths (e.g., 5, 10, 20, 40, and 80 cm or 15, 30, 60, and 120 cm. In these cases, the installation depth of the topmost sensor (e.g., 5 cm) is always multiplied by 2: $a(n) = 5$ cm $* 2 (n - 1)$, where a = installation depth (cm) and n = number of steps. It has to be noted, that these depths are only a generalized suggestion and actual installation depths *should be always adapted to local site conditions and the experimental question*. For some sensors (e.g., soil water content sensors) the presence of soil horizons within the soil profile has to be also taken into account because sensor installation at the interfaces between soil horizons will influence measured data and later cause serious problems interpreting the data so collected.

1.4.1.2. Installation of sensors in space

1.4.1.2.1. Arrangement of replicates

As Ghodrati & Jury (1990) pointed out, diverse arrangements of instruments may lead to very different interpretations of the state distribution, and thus affect the interpretation of the processes. Hence, some attention must be devoted to the spatial arrangement of the sensors in the experimental setup. If, for example, three instruments of the same type are going to be installed at a given depth as replicates, it should be avoided that all devices for one depth are clustered and the cluster at the next depth level is at close distance (see Fig. 1.4).

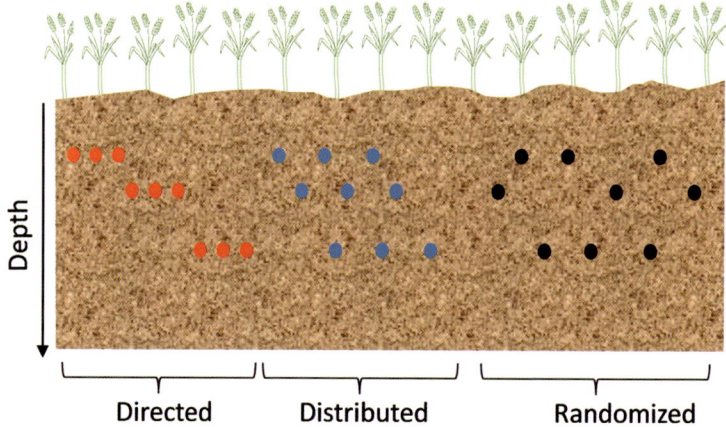

Fig. 1.4. Possible arrangements of 9 sensors with 3 replicates in 3 installation depths.

In soils where soil properties change greatly at short horizontal distances (e.g., changes in horizon depths, surface structures such as tractor tracks) the directed sensor pattern may result in different mean values for each sampling depth compared to values obtained by using a distributed or the randomized installation pattern (Fig. 1.4). In extreme cases, the directed pattern may lead to false conclusions. Additionally, soil heterogeneity, mutual influence, and other factors must be considered in order to avoid systematic errors caused by sensors and by the installation procedure of the sampling devices in the field. A randomly distributed sensor- or sampling system was described e.g. by Wessel-Bothe (2002) who used such experimental design to study the migration of solutes through the soil.

Generally, if sensors are to be installed at different depths, sensors or sampling systems should never be installed below each other to rule out shading effects. Additionally, the use of some sampling systems (e.g., soil water extraction systems) will invariably influence the natural water and solute flow locally. For this reason, sensors should always be placed at a certain distance from each other in order to avoid artifacts by neighboring instruments.

1.4.1.2.2. Installation from the surface or from a profile wall?

After a decision has been made with respect to distribution of sensors or sampling devices with depth and in space for a particular site, it must be decided *how* the devices will be installed in the field. In general, there are two different options: i) installation from the soil surface or ii) installation from a pit or trench. Irrespective of the installation method of choice (from the surface or from a pit/trench), any hydrological shortcuts have to be avoided. As mentioned, different ways of sensor/sampling device installation procedures are feasible.

A) Diagonal sensor installation from the surface

The instrument is installed from the soil surface at a given angle, e.g. 45°, to avoid direct flow of precipitation along the shaft to the measuring- or sampling tip of the device. It has to be noted, that surface installation is always prone to the risk of hydraulic shortcuts in dry soils (with a potential to shrink) or if the diameter of the borehole for sensor installation is larger than the sensor shaft itself. To install the sensor, the borehole is deepened at a constant angle, whereby the desired length of the installation borehole must be calculated trigonometrically and depends on the angle and the desired installation depth: $L = D/\sin \alpha$, where L = length of the installation borehole, D = desired depth of installation and α is the angle of the installation tool relative to the soil surface. For example, if the desired installation depth is 50 cm below surface and the installation angle is 45°, the borehole must have a length of ~71 cm.

B) Installation from a soil pit or trench

Many scientists prefer the installation of sensors/devices from a profile pit or trench to avoid the risk of hydraulic shortcuts by vertical installation. Another advantage of the horizontal installation from a pit/trench is that the sensors/sampling systems do only provide measurements from one horizontal depth. This holds especially for water content sensors, which classically integrate their measurement over a specific sensor length (see chapter 4).

Whenever horizontal installation of sensors is carried out, the influence on the state variables in the undisturbed soil profile from the soil pit/trench must be considered, even if the pit/trench is backfilled after installation. The reason for this is that the native soil structure of the excavated soil material is destroyed by the excavation process, homogenizing the excavated material. This invariably causes differences in soil physicochemical properties and water, solute, air, as well as thermal fluxes. These modifications will not only change the soil state parameters within the backfilled soil but also in the adjacent natural soil to which it is connected lateral by (e.g., hydraulic) exchange and other processes. Consequently, sensors/sampling systems installed from a pit/trench should be installed in the native soil at a certain distance (classically 30 cm are recommended) from the pit/trench wall in order to minimize their influence on the measurements.

That an insufficiently compacted pit/trench will begin to settle and compact naturally over time due to the overburden and changes in water content is obvious and must be another concern. Compaction has the effect that a constant, natural bulk density and soil structure is only reached after several months to even years. Because the installed instruments are installed within the static natural soil (the trench wall) and cable and tubes are run through the backfilled soil material, compaction of the refill may cause stress (pull) on the cables/tubes. In the worst case, this stress may propagate to the sensor/sampling device and may cause the loss of contact of the sensor/sampling device and the native soil surrounding it and cause instrument malfunction or failure.

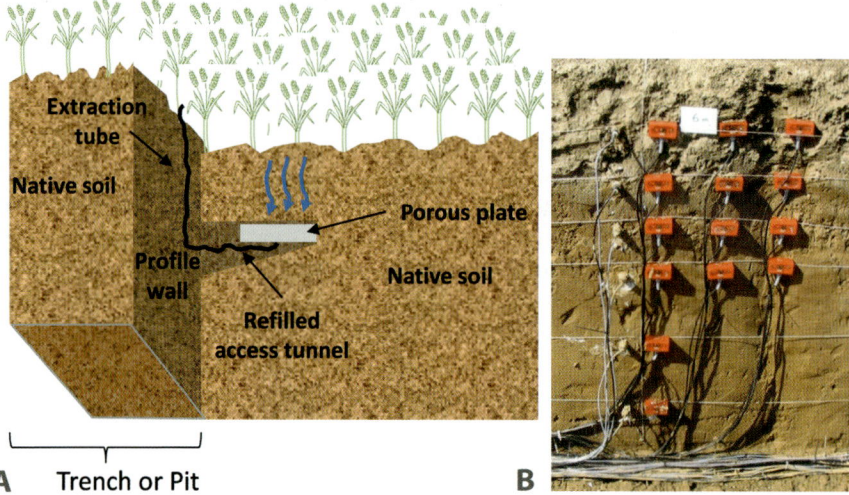

Fig. 1.5. Installation of instruments from a profile pit. A) schematic sketch of the installation of a porous suction plate and b) photo of installed suction cups, tensiometers, and TDR-sensors (photo: Lutz Weihermüller).

To reduce the buildup of such stresses, such pits/trenches should be backfilled layer by layer (with a layer thicknesses of 10–15 cm each) and each layer should be compacted manually or by appropriate machines. Ideally, the soils from different horizons should be separated while the pit/trench is being dug and refilled according to its natural stratigraphy. During refilling, care must be taken not to disturb exposed cables and tubes. To this end, it may be very useful to fix cables and tubes to the profile wall with e.g., tent pegs.

C) Installation of sensors without shaft from the surface and refilling the borehole

Some devices have no extensions or shafts, so only power supply and data transfer cables or vacuum and extraction tubes must be run from the instrument to the surface. In these cases, the borehole above the sensor should be filled with slurry made of the parent material dug out for installation. The risk of causing hydrological shortcuts may be avoided by thoroughly filling the borehole partly or completely with bentonite pellets or a mixture of bentonite and soil slurry. In the latter case, it has to be made sure that the bentonite does not affect local water contents to such an extent that the bentonite influences the measurements (e.g., water content).

As bentonite will shrink under very dry conditions, this is however only working under the condition that the soil water content remains above a minimum which prevents the formation of cracks and thus hydraulic short cuts. As Wessel-Bothe (2002) showed, the breakthrough of solutes over a period of 22 months was not significantly affected by the vertical installation of shaftless suction cups even if the boreholes were completely filled with parent soil material and without bentonite.

1.5. Arrangement of enclosures, cabinets, housings etc.

All sensors require dedicated data loggers and power supplies and to operate water extraction systems, sampling bottles and vacuum pumps must be installed also. In most cases, these peripheral devices are located in enclosures, housings, or cabinets to protect sensible electronical parts and sampling bottles from precipitation, light, and/or temperature. The enclosures, cabinets, or housings will shadow the rain in direct vicinity, on the one hand, and cause local preferential infiltration paths where the water runs off the enclosure, housing, or cabinet. Therefore, these above-ground components should be set up as far away as possible from the buried sensors. On the other hand, the distance between the sensors/devices and the loggers, power supply, pumps, or sampling bottles should be kept as short as possible in order to reduce the risk of damage (e.g., by animals, machines) and dead volume (for water extraction devices). In some cases, long cables may unduly attenuate the measurement signal, especially weak analog signals.

After the instruments are installed, avoid any artificial disturbance of the measurement plot e.g. by treading on it or vehicular traffic.

1.6. Post-processing of acquired data

After the measurements were performed, appropriate data post-processing must follow. For detailed information about appropriate statistical analysis we refer to the respective literature (e.g., Sokal and Rohlf 2012) and here, we only restrict our recommendation to two major often observed problems.

1.6.1. Calculating mean concentrations from multiple, measured concentrations

All solute extraction devices (see chapter 6) produce two types of data, namely the volume of extracted soil pore water and the concentration of the target substance in it. If more than one sampler in one location (e.g., at different depths) was installed, a mean concentration is often calculated from the samplers as the first step of data post-processing. Due to differences in the amount of soil pore water extracted by each individual sampler device, the mean of all samples cannot directly be calculated from the different measured concentrations. Instead, the solute masses and water amounts of each single measurement should be summed up, and from those sums the bulk or mean concentration should be calculated. Alternatively, the replicate samples of a given depth level may all be lumped together and the concentration of this sample will provide the mean concentration. This procedure of course precludes the determination of a standard deviation at this given depth level from the corresponding device data and is unable to supply information on the variation of concentrations at that depth level.

1.6.2. Calculating mean pF and pH-values from replicate measurements

Because matric potential and proton (H^+) concentrations are classically expressed in logarithmic form (pF and pH-value), the calculation of mean values is not straight forward. Calculating the mean value of measured pF and pH values might cause wrong mean values. Therefore, the data have to be converted to absolute (e.g., H^+-)concentrations, from which a mean can then be computed. Finally, the mean pF and the mean pH values may be calculated from the mean of the absolute values.

1.7. Concluding remarks

The authors did not intend to be complete in listing and describing all existing sensor types and measurement techniques. Moreover, the most accepted sensors/measurement techniques will be described and discussed. Also some novel sensors/measurement systems are associated to brand names and sensors of similar type are also often available and can be used instead.

1.8. References

Bear, J., 1972. Dynamics of Fluids in Porous Media. Elsevier, New York

Durner, W., Flühler, H., 2005. Soil hydraulic properties. In: Anderson, M.G. (Ed.), Encyclopedia of Hydrological Sciences. John Wiley & Sons.

Ghodrati, M., Jury, W.A., 1990. A field study using dyes to characterize preferential flow of water. Soil Sci. Soc. Am. J. 54: 1558–1563.

Sokal, R.R., Rohlf, F.J., 2012. Biometry – The Principles and Practice of Statistics in Biological Research. 4th Edition. W.H. Freeman and Company, New York.

Weihermüller, L., Kasteel, R., Vereecken, H., 2006. Soil Heterogeneity Effects on Solute Breakthrough Sampled with Suction Cups: Numerical Simulations. Vadose Zone Journal 5 (3): 886–893, doi:10.2136/vzj2005.0105

Wessel-Bothe, S., 2002: Simultaner Transport von Ionen unterschiedlicher Matrixaffinität in Böden aus Löss unter Freilandbedingungen – Messung und Simulation. Bonner Bodenkundl. Abh., Bd. 38, University of Bonn, Germany.

Zurmühl, T., 1994. Validierung konvektiv-dispersiver Modelle zur Berechnung des instationären Stofftransports in ungestörten Böden. Ph.D thesis, Hydrology, University Bayreuth, Germany.

Soil physicochemical parameters

2. Soil redox potential

T. Mansfeldt

2.1. Introduction

2.1.1. Objectives of redox potential measurements in soils

During oxidation and reduction reactions, electrons are transferred form one chemical element (in its elemental form, as ion or molecule) to another element. Because free electrons do not exist in chemical reactions, oxidation, i.e. the loss (donation) of an electron, and reduction, i.e. the gain (acceptance) of an electron, are always coupled. By the transfer of electrons, energy is additionally transferred which is the energy of life. Hence, reduction–oxidation (redox) reactions support the life on Earth, even in soils.

The most important source of electrons in soils is reduced carbon (C) occurring in the soil organic matter (SOM) pool. By photosynthesis, tetravalent oxidized C in carbon dioxide (in CO_2, C has an oxidation state of +IV) is reduced to organic C species (with an oxidation state of 0) and plant biomass is formed (Eq. 2.1):

$$6CO_2 + 6H_2O = C_6H_{12}O_6 + 6O_2 \qquad (2.1)$$

The SOM pool is continuously replenished by inputs of dead plant and animal residues and can be considered a large electron reservoir (electron donor). Metabolizing plant roots and microorganisms are able to oxidize the reduced C forms enzymatically. Electrons released during the oxidation of C are transferred to elemental oxygen (which occurs as oxygen gas O_2) which, in turn, is reduced to water (H_2O). Hence, O_2 is the terminal electron acceptor. Soil redox conditions under which O_2 is stable and hence available are called *oxidizing*, and characterized by low electron availability. The O_2 pool of soils is continuously replenished by O_2 diffusion through soil pores, as long as they are filled with air. When filled with water (caused by high ground water levels, perched water table, natural or artificial flooding), O_2 diffusion is extremely slow and depending on metabolic activity, the soil O_2 pool is more or less rapidly exhausted. Soil redox conditions under which O_2 partial pressure is low or O_2 is absent are called *reducing*, and are characterized by high electron availability. Under reducing soil conditions, elements other than O are the terminal electron acceptors. These include pentavalent nitrogen (N^V) in nitrate (NO_3^-), tri- and tetravalent manganese ($Mn^{III, IV}$) in Mn oxides (e.g., birnessite, δ-MnO_2), trivalent iron (Fe^{III}) in Fe oxides (e.g., goethite, α-FeOOH), hexavalent sulfur (S^{VI}) in sulfate (SO_4^{2-}) and tetravalent C in CO_2. Although, some overlap may occur, the use of the different electron acceptors is a stepwise one and known as the sequential reduction sequence (Ottow 2011, Ponnamperuma 1972).

Redox conditions of soils are classically assessed by measuring the redox potential, which is abbreviated either as Eh or E_H. The letter E denotes the electrode potential and h/H represents the element hydrogen (H). As symbols of chemical elements are generally capitalized, the abbreviation E_H is used here. Sometimes the abbreviation ORP (Oxidation–Reduction Potential) is used.

According to the preferential electron acceptor, a more precise classification of the soil redox status (at pH 7) can be defined (Reddy and DeLaune 2008, Zhi-Guang 1985): *oxidizing* conditions prevail where the E_H is > 300 mV and O_2 is predominant; *weakly reducing* conditions occur where E_H ranges from 300 to 100 mV and NO_3^- and $Mn^{III,IV}$ are reduced; *moderately reducing* conditions are where E_H ranges from 100 to –100 mV and Fe^{III} is reduced; under *strongly reducing* conditions SO_4^{2-} and CO_2 are reduced and the corresponding E_H is < –100 mV. Figure 2.1 illustrates this sequential reduction sequence. It is worth noting that some redox species change aggregate state during the electron transfer, e.g., from an aqueous species (NO_3^-) to a gas (N_2).

Although, subject to some limitations, redox conditions may be assessed by measuring the soil redox potential with a platinum (Pt) electrode. The assessment of the soil redox potential is particularly useful to characterize the onset of reducing condition in soils caused by a lack of O_2

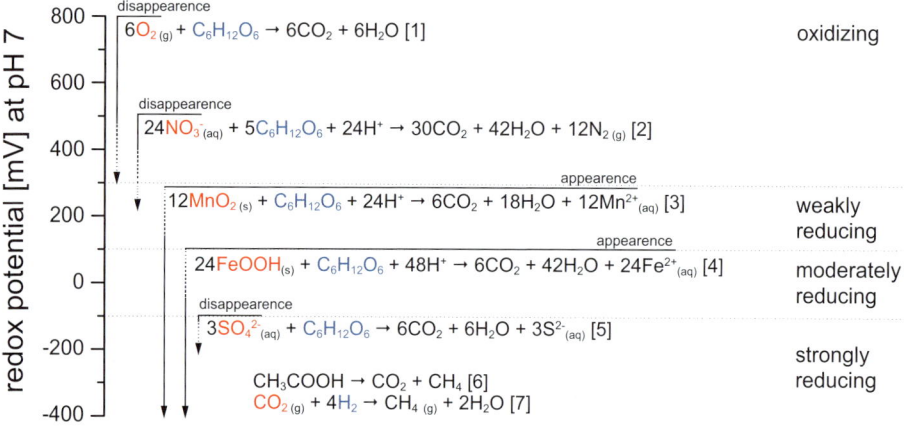

Fig. 2.1. The idealized sequential reduction sequence in waterlogged soils and a classification of redox zones. The primary electron donor (denoted in blue) is the soil organic matter, which is designated for simplicity by the glucose molecule. The different terminal electron acceptors are denoted in red. Note that there are no sharp boundaries for the redox processes but some overlaps. Aggregate states are aqueous (aq), gaseous (g), and solid (s).

and associated biogeochemical processes such as denitrification and redoximorphosis (creation of distinctive soil color pattern). Some detailed examples will be given in section 3.5.2 and important restrictions for the interpretation of measured redox potentials in section 3.5.3.

2.1.2. Fundamentals of redox reactions

In the following section, the theoretical background of the redox potential and the role of protons in redox reactions will be outlined. For a more in-depth discussion of these issues, the reader is referred to textbooks, e.g., Essington (2015) and Strawn et al. (2015).

The theory behind redox potential can be derived by considering the general redox half reaction (in this case the reducing reaction)

$$\text{oxidized species} + m\text{H}^+ + n e^- = \text{reduced species} \qquad (2.2)$$

where m is the number of protons (H^+), and n is the number of electrons (e^-) participating in the reaction. This reaction is expressed quantitatively by calculating the Gibbs free energy (ΔG):

$$\Delta G = \Delta G^0 + RT \ln \frac{(red)}{(ox)(H^+)} \quad (2.3)$$

where ΔG^0 is the standard free energy change, R the universal gas constant (8.3145 j mol^{-1} K^{-1}), T the absolute temperature (298.15 K), and round parentheses denote the activities of the oxidized (ox, e.g., Fe^{3+}) and reduced (red, e.g., Fe^{2+}) species (mol l^{-1}). By converting Gibbs free energy into voltage, the relation to the redox potential becomes obvious:

$$E_H = E^0 - \frac{RT}{nF} \ln \frac{(red)}{(ox)} - \frac{mRT}{nF} \ln(H^+) \quad (2.4)$$

where E_H is the electrode potential (redox potential) for the reaction (V), E^0 the standard electrode potential (V), and F the faraday constant (96,485 C mol^{-1}). Equation 2.4 is the Nernst equation. Substituting the values of R, T, and F and using the relationship $\ln(x) = 2.303 \log(x)$ and substituting pH for $-\log(H^+)$, simplifies Eq. 2.4 to:

$$E_H = E^0 - \frac{0.059}{n} \log \frac{(red)}{(ox)} + \frac{0.059m}{n} pH \quad (2.5)$$

Alternatively, the redox potential of a reaction can be expressed in units of pe, a constant that is derived from the equilibrium constant of a redox reaction and which considers information on pH. The pe value can be viewed as a measure of the electron activity in the system (Lindsay 1979) and is calculated from the redox potential:

$$E_H = \frac{2.303RT}{nF} pe \quad (2.6)$$

where pe is the negative common logarithm of the electron activity (pe = $-\log(e^-)$). For $n = 1$ and $T = 25$ °C, Eq. 2.6 simplifies to

$$pe = \frac{E_H}{0.059} \quad (2.7)$$

where E_H is given in Volts (V).

Furthermore, plots of pe+pH may be used to show how redox potential affects the solubility of elements under consideration. As Lindsay (1979) pointed out, this redox parameter partitions the H$^+$ ions of an overall chemical reaction into those associated with the redox component of the reaction from those associated with the acid-base component.

In practice, redox potentials are determined by measuring an electrical potential between two electrochemical half-cells. Each half-cell contains an aqueous solution of a distinct redox pair and a Pt electrode that functions as a sink or source of electrons to the solution. When the half-cells are connected, an electrochemical cell is created and electrons start to flow. A switch prevents the flow of electrons, and the potential between these two half-cells is measured using a voltmeter. During this process, the voltmeter registers the difference in electrical potential between the two cells. When one of the electrodes is a standard hydrogen electrode (SHE), and the other cell contains a redox pair whose activity ratio of reduced/oxidized species is equal to one and equilibrium in electron transfer is attained, the measured redox potential is the standard electrode potential or standard-state reduction potential. By convention, the SHE consists of a Pt electrode which dips into a solution containing H$^+$ ions at unit activity (1 M, that is pH = 0) and hydrogen (H_2) gas moving across the surface of the Pt surface at 0.101 MPa pressure and a temperature of 25 °C (298.15 K). The reversible half reaction of the SHE is:

$$H_2 = 2H^+ + 2e^- \qquad E^0 = 0 \text{ V} \qquad (2.8)$$

As a reference point for all other electrode reactions, the SHE's potential (E^0) is defined to be zero. The SHE is defined as the standard reference and the potential of any other redox pair can be measured against this using a second electrode. When the second electrode, for example, contains the Fe^{3+}–Fe^{2+} redox pair at unit activity and at equilibrium, the standard potential of this reaction is +0.77 V

$$Fe^{3+} + e^- = Fe^{2+} \qquad E^0 = 0.77 \text{ V} \qquad (2.9)$$

The overall redox reaction can be written as

$$2Fe^{3+} + H_2 = 2Fe^{2+} + 2H^+ \qquad (2.10)$$

The relative order of oxidation and reduction of redox pairs relative to each other can be predicted by comparing the redox pairs' standard redox potentials. The redox pair with the higher standard potential, e.g., Fe^{3+}–Fe^{2+}, will reduce and oxidize at a higher redox potential than the redox pair with the lower standard potential, e.g., S^0–S^{2-}.

Table 2.1 presents standard-state potentials of some redox pairs of interest in soils.

Table 2.1. Standard-state reduction potentials (E^0) and corresponding potentials at pH 7 (E^0_7) of selected redox pairs of significance in soils (according to Reddy and DeLaune 2008).

Reduction Reaction	E^0 (mV)	E^0_7 (mV)
$O_2 + 4H^+ + 4e^- = 2H_2O$	1230	810
$2NO_3^- + 12H^+ + 10e^- = N_2 + 6H_2O$	1240	820
$MnO_2 + 4H^+ + 2e^- = Mn^{2+} + 2H_2O$	1290	870
$Fe(OH)_3 + 3H^+ + e^- = Fe^{2+} + 3H_2O$	800	380
$SO_4^{2-} + 10H^+ + 8e^- = H_2S + 4H_2O$	340	−80
$CO_2 + 8H^+ + 8e^- = CH_4 + 2H_2O$	170	−250
$2H^+ + 2e^- = H_2$	0	−420

The relationship between redox potential and pH becomes obvious by considering the Nernst equation Eq. 2.4, if no protons are involved in the half-reaction, the pH term is zero) and the chemical reactions in which protons are involved (Eq. 2.2). In soils, nearly all redox reactions involve the transfer of protons and hence redox reactions cause the soil's pH to change. Under reducing conditions, protons are typically consumed and the pH increases, whereas under oxidizing conditions, protons are produced and pH drops. At equal number of protons and electrons, i.e. $mH^+ = ne^-$ or $m/n = 1$, for each increase in pH unit, the corresponding redox potential decreases by 59 mV (the so-called Nernst factor). In the biological sciences, redox potentials are often reported relative to pH 7 because plant and animal cells have a pH around 7. Such pH-conversion is frequently done for soils and sediments, because the standard potential is related to pH 0, which is far from natural conditions. Converting field soil redox measurements into values corresponding to pH 7 (donated as E_{H7}) can be achieved by applying Eq. 2.11 to the redox potentials measured in the field:

$$E_{H7} = E_H + (pH - 7) \cdot 59 \qquad (2.11)$$

where E_H is the measured redox potential expressed in millivolts and pH is soil's pH. Although, the value of –59 mV per pH unit is reasonable for most measurements in soils (Bohn et al. 2001), the m/n ratio of many redox reactions is greater than 1. For example, m/n is 2 for the MnO_2–Mn^{2+} pair and m/n is 3 for the $FeOOH$–Fe^{2+} pair (Table 2.2). Because many redox couples (including organic couples with an unknown ratio of m/n) contribute to the redox potentials of soils, the conversion approach and the use of the Nernst factor in general has been criticized (Bohn et al. 2001). In any case, whenever a soil redox potential is measured, the pH should also be recorded and both data should be reported.

Table 2.2. Examples of redox half reactions showing the effect of protons (H^+) on redox potential (E_H). Number of protons are given by m, and number of electrons (e^-) are given by n (according to Reddy and DeLaune 2008, modified).

Reduction Reaction	m/n	E_H/pH (mV/pH unit)
$O_2 + 4H^+ + 4e^- = 2H_2O$	1	59
$NO_3^- + 2H^+ + 2e^- = NO_2^- + H_2O$	1	59
$2NO_3^- + 12H^+ + 10e^- = N_2 + 6H_2O$	1.2	78
$MnO_2 + 4H^+ + 2e^- = Mn^{2+} + 2H_2O$	2	118
$FeOOH + 3H^+ + e^- = Fe^{2+} + 2H_2O$	3	177
$Fe_3(OH)_8 + 8H^+ + 2e^- = 3Fe^{2+} + 8H_2O$	4	236

Note, that the standard-state potentials listed in Table 2.1 can be used in Equation 2.5 to calculate standard-state reduction potentials related to pH 7, i.e. E^0_7, which are also listed in Table 2.1.

2.2. Method selection

Several methods are available to identify and sometimes to monitor reducing conditions in the field. One simple method to infer reducing conditions is the use of dyes, which react with Fe^{2+} to form coloured compounds. Commonly, 2,2'-dipyridyl is used (Childs 1981). Although, simple to use, the dye method provides only a snapshot, and it uses a harmful chemical. Another but much more complex approach to predicting dominant redox reactions is to simultaneously measure the concentrations of electron acceptors (e.g., dissolved O_2, NO_3^-, SO_4^{2-}), intermediate products (dissolved hydrogen, H_2), and final products of redox reactions (e.g., Fe^{2+}, dissolved hydrogen sulfide, H_2S) along a redox gradient. This approach was introduced by Chapelle et al. (1995) for groundwater systems and is called terminal electron-accepting processes (TEAPs). It provides a better understanding of redox processes in groundwater systems compared to E_H-based methods. It has been applied to the pore waters of Californian paddy soils (Gao et al. 2002). However, TEAPs have disadvantages for pedological systems, where the focus of interest, compared to hydrological systems, is in a smaller vertical and horizontal scale (Chapelle et al. 1995). In addition, identifying dominant TEAPs requires intensive data collection and chemical analysis (Gao et al. 2002), rendering it unsuitable as a simple field method. Scott and Morgan (1990) established the concept of oxidative capacity (OXC) to describe redox conditions in sediments via geochemical classes. Oxidative capacity integrates all the major oxidized and reduced species into a single conservative parameter (e.g., oxic or postoxic) after defining electron reference levels (e.g., HS^-). In OXC computations, besides water analyses, soil analyses are also required to determine the oxidative capacities of Mn and Fe oxides. However, no reliable methods to estimate soil Mn and Fe oxides concentrations as electron acceptors (bioavailable concentration) exist, which restricts the use of the OXC approach (Gao et al. 2002). Other tools to quantify or qualify the oxidation-reduction status of soils involve the use of striated polymer plates coated with synthetic ferrihydrite (Fakih et al. 2008), or Fe metal rods to infer the O_2 concentration in and the aerobic status of a soil (Owens et al. 2008). Whether a soil is in a reduced state or not can be checked by the Indicator of Reduction in Soils (IRIS) method, introduced by Jenkinson and Franzmeier (2006) and adapted by Rabenhorst and Burch (2006) and Castenson and Rabenhorst (2006). Synthetic Fe oxides are coated on polyvinylchloride (PVC) tubes, which are installed for a distinct period of time in the soil. If O_2 is depleted due to water saturation, the Fe^{III} at the PVC tube is reduced to the soluble Fe^{2+}. After the tubes are removed from the soil, they can be visually assessed for the effects of reduction by the depletion of the oxide-coating. Meanwhile, the IRIS device is, within the framework of the US Hydric Soil Technical Standard, a method of proving reducing soil conditions (NTCHS 2007). Quite recently, Stiles et al. (2010) introduced PVC tubes coated with synthetic Mn oxides, called the Manganese Indicator of Reduction in Soils (MIRIS). Dorau and Mansfeldt (2015) and Dorau et al. (2016) improved this method and recommend it for short-term monitoring because tri- and tetravalent Mn is the preferred electron acceptor compared to trivalent Fe, and this offers the possibility of additionally distinguishing between weakly and moderately reducing conditions. Although, this is a simple and cheap method (IRIS/MIRIS), synthesizing the oxides and applying them onto the PVC bars requires a lot of skill and laboratory equipment.

The method presented below using Pt electrodes for redox potential measurements is relatively easy to perform and allow the redox state to be continuously monitored via data logging. Although, the Pt electrodes are quite expensive and so is the data logger, and a certain number of electrodes is required to achieve reliable results due to soil heterogeneity, many researchers applied this method in the field. Measurement of the redox potential using the Pt electrode was first initiated in the 1920s (Gillespie 1920) and since then many studies about *in-situ* measurements in soils have been published (see references cited later). Furthermore, the Pt electrode has been used also in the laboratory in soil columns (Rezanezhad et al. 2014, Weigand et al. 2010, Whisler et al. 1974) and soil suspensions (Patrick 1966, Rennert and Mansfeldt 2005). Furthermore, the Pt electrode was applied to O_2-deficient environments like estuary (Hinchey and Schaffner 2005), marine and brackish sediments (Gerwing et al. 2013, Plante et al. 1989), lake sediments (Olsen and Andersen 1994) and constructed wetlands (Dusek et al. 2008, Kleimeier et al. 2014). Hence, the following instructions to construct and use single Pt-tipped electrodes may be useful for not only soil scientists, but also for researchers dealing with other natural or anthropogenic reducing environments. Because redox potentials often reveal a large spatial variability, efforts have been made to build redox probes that consists of multiple Pt electrodes (Teasdale et al. 1998, Vorenhout et al. 2011, Vorenhout et al. 2004). However, these will not be considered in the following text.

2.3. Method application

2.3.1. Equipment

To determine the soil redox status in the field, the following equipment is required: (i) Pt electrode, (ii) reference electrode, (iii) salt bridge, and (iv) a portable voltmeter. Figure 2.2 shows a sketch of such an arrangement installed in a soil and its basic mode of operation.

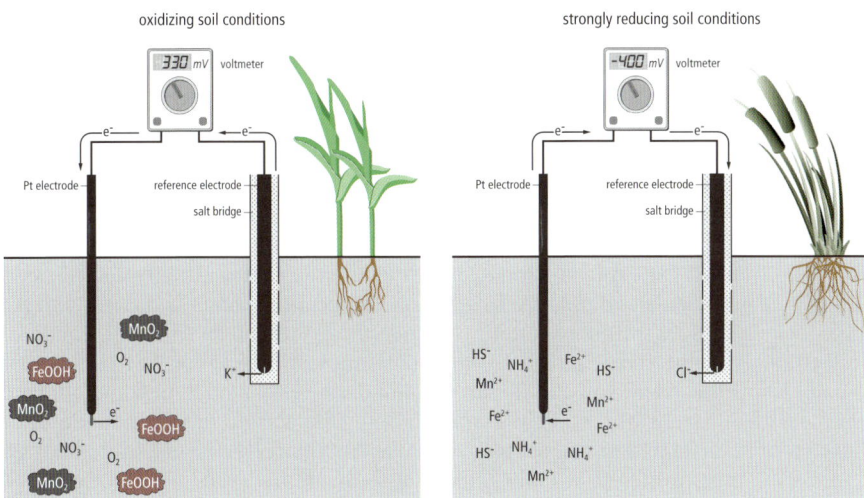

Fig. 2.2. Sketch of the design for measuring of the redox potential in soils. To calculate the redox potential, 207 mV (at 25 °C) have to be added to the voltmeter readings because in the field not the standard hydrogen electrode is used but a reference electrode (in this example a silver/silver chloride electrode with an electrolyte concentration of 3 M KCl whose standard state redox potential is below 207 mV against the standard hydrogen electrode).

Instead of using a portable voltmeter, readings can also be done automatically with a data logger. In the recent past, the latter has becoming more and more popular.

2.3.1.1. Platinum electrode

Redox electrodes or probes are typically constructed by soldering Pt wire or sheet to a copper (Cu) wire or rod, sometimes brass is used. These electrodes can be called Pt-tipped redox electrodes. Platinum is used because it is non-selective, meaning it will accept electrons from all redox reactions, and it is inert, meaning it will accept or release electrons but will not react with any species in the solution. Principally, other noble metals like gold (Heduit et al. 1993), wax impregnated graphite (Biddle et al. 1995) or glassy carbon (Teasdale et al. 1998) can be used but this is not common. Properly working Pt electrodes have a Pt purity that ranged from 98.95% (Wafer et al. 2004) to 99.998% (Rabenhorst et al. 2009). Of outstanding importance is the construction of the junction Pt-Cu wire/brass rod, which must be waterproof (and resistant to microbial degradation, compatible to plastic, electrically insulating), especially when used in soils which are temporarily or permanently water-saturated. If water encounters the junction Pt-Cu wire/brass rod by micro cracks, an electrical shortcut arises, which manifests itself in strongly oscillating readings. Such electrodes are unusable, must be removed from the soil and repaired.

Various construction methods have been used during the past 50 years (see Wafer et al. (2004) and literature cited therein). However, the best way of waterproofing the electrode is to sinter the Pt-Cu wire junction by a ceramic jacket as recommended by Pfisterer and Gribbohm (1989). Own experience has shown that these Pt electrodes are reliable and long lasting even under water-saturated conditions (Mansfeldt 1993, 2003). After being in the field for up to 48 months, none of 25 ceramic-jacket electrodes was faulty. In a running study, commercially available electrodes using a similar stable junction installed in a marsh soil have been working properly and fault free for a period of 72 months.

In the following, the author gives a description for the construction of a Pt electrode (Fig. 2.3a). Before preparing, Pt wires (purity of Pt $\geq$99.5%, Pt sheets can also be used) used should be stored overnight in a solution of concentrated nitric acid (HNO_3) in order to remove any surface impurities. After washing with demineralized water, the Pt wire (2 mm diameter, 20 mm length) is welded with gold solder onto a Cu wire (2 mm diameter, 40 mm length). The Cu wire is connected to a Cu lead (length depending on the length of the electrode body plus either about 3 cm when the voltmeter is directly at the electrode connected or several meters when a central measuring container is used), the insulation of which was pushed back several centimeters by means of a Cu tube. A crimping tool may be used to squeeze together the Cu tube. The section from the weld to the Cu lead is coated with a ceramic jacket at 880 °C. This step can be performed in, e.g., a dentist laboratory. In addition, the ceramic jacket is coated with plastic. The ceramic jacket excludes intruding water even under long-term water saturated conditions as ceramic is extremely stable. The Cu lead's insulation is moved back down to the Cu tube. The electrode body is introduced into an acrylic tube (8 mm diameter, length depending on the depth of measurement) leaving 10 mm of Pt wire protruding. The acrylic tube is completely filled with a water proof resin using a syringe with a cannula. Alternatively, the electrode body can be enclosed in a PVC pipe (Wafer et al. 2004), fiberglass (Swerhone et al. 1999), or carbon fiber (Vorenhout et al. 2004).

It is important that the exposed Pt surface area ('active surface') of a set of redox probes that are designed for the same measuring campaign have similar active surfaces because the contact area soil/Pt surface influences the redox potential. The active surface area of the above presented electrodes amounted to about 125 mm^2. Teichert et al. (2000) used electrodes with an active surface of 30 to 35 mm^2 and obtained reliable results in a sandy forest soil.

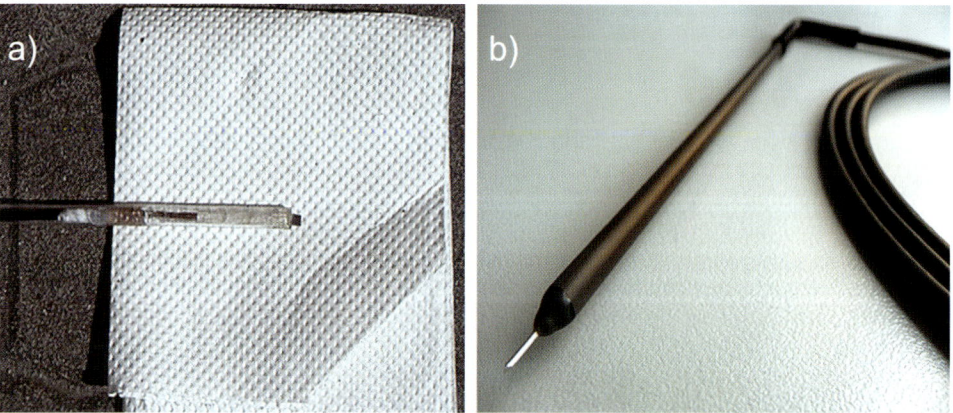

Fig. 2.3. A homemade (a) and a commercially available (b) platinum electrode for the determination of the redox potential in soils. The tip is either a platinum sheet (a) or a platinum wire (b).

Platinum electrodes (both single-tip and multiple) specifically designed for measuring the redox potential under field conditions are also commercially available (Fig. 2.3b).

After manufacturing, the redox electrodes should be tested in redox buffer solutions to check if they are working properly. Especially, when using Pt electrodes whose Pt-Cu wire junction are epoxy-sealed, this check is very important because holes in the epoxy or incomplete coverage of the epoxy can leave exposed Cu. When installed in the field, an electrical connection of Cu with the soil solution occurs. Redox buffer solutions contain a redox pair, e.g., Fe^{3+}–Fe^{2+} at high concentration, and hence, are highly buffered (in the older literature termed as 'poised') systems. Accordingly, any electrode inserted in such buffer solutions should indicate (i) the exact value (±5 mV) of the buffer solution and (ii) there should be no drift in the reading. Table 2.3 presents commonly used redox buffer solutions. Some of them are commercially available but they may easily be prepared in the laboratory. Owens et al. (2005) presented a detailed study on the pre-evaluation of Pt electrodes before field-installation using different redox-buffer solutions. They stated, that quinhydrone at pH 7, which is commercially available, is not a rigorous standard for determining improper function of Pt electrodes utilizing Cu wire (as it is usually the case).

Table 2.3. Reference solutions used to test redox electrodes. [a] SHE, standard hydrogen electrode.

Reference Solution	Potential (mV, vs SHE[a] at 25 °C)	Reference
ZoBell's solution 3.3×10^{-3} M $K_3[Fe^{III}(CN)_6]$ + 3.3×10^{-3} M $K_4[Fe^{II}(CN)_6]$ in 0.1 M KCl	+430	ZoBell (1946)
15% (w/v) $Ti^{III}Cl_3$ in 0.2 M sodium citrate	–480	Zehnder and Wuhrmann (1976)
Saturated quinhydrone in 0.05 M potassium biphthalate (pH = 4.008)	+462	Dirasian (1968)
M $Fe^{II}(NH_4)_2(SO_4)_2 \cdot 6H_2O$ + M $Fe^{III}NH_4(SO_4)_2 \cdot 12H_2O$ in 1 M H_2SO_4	+675	Light (1972)

Often, the Pt electrodes described above are referred to as microelectrodes. Typically, they have an active surface at the mm^2-scale. For some time, soil redox electrodes at the nm^2-scale have been developed (Jang et al. 2005, Pang and Zang 1998). It seems to be reasonable to restrict the term 'micro' solely to the latter one.

2.3.1.2. Reference electrode

As pointed out, the reference electrode for measuring redox voltages is the SHE (see section 1.2). In the field, however, it is not trivial to operate and maintain a SHE. Instead, either a silver/silver chloride (Ag/AgCl) electrode or a calomel (Hg/Hg_2Cl_2) electrode is used as a reference. Although, properly working, the use of a calomel electrode in the field is not recommended because Hg is poisonous and its use should be avoided. Thus, in the following the use of the Ag/AgCl-electrode is described.

An Ag/AgCl electrode consists of an Ag wire surrounded by solid AgCl, contained in a solution of KCl (1 mol l^{-1}, 3 mol l^{-1}, or saturated). Chloride (Cl^-) is added because it forms sparingly soluble silver chloride (AgCl) with Ag. The activity of the silver ions (Ag^+) is determined by the solubility product of silver chloride on the activity of chloride ions. The potential-determining step is the oxidation of Ag, a reversible redox half-reaction (Eq. 2.12), and the subsequent precipitation of silver chloride (Eq. 2.13):

$$Ag^0 = Ag^+ + e^- \qquad (2.12)$$

$$Ag^+ + Cl^- = AgCl \qquad (2.13)$$

Connecting the Pt electrode and the reference electrode to the voltmeter, closes the electrical circuit and electrons start to flow. If the soil is under oxidizing conditions, the reaction will go to the right hand side (Ag is oxidized) and the voltmeter will record a positive flow of electrons (Fig. 2.2). If instead the Pt electrode is applied in a reducing environment, electrons will flow towards the reference electrode and the reaction (Eq. 2.12) will proceed to the left hand side (Ag is reduced). The voltmeter will record a negative flow of electrons. Charge balance is maintained within the reference electrode by diffusion of ions through a porous ceramic diaphragm or membrane tip. Potassium ions (K^+) will migrate across the diaphragm into the soil when the redox potential is positive, and chloride ions will migrate at negative redox potentials (Fig. 2.2). The voltage difference between the Pt and the reference electrode must be corrected for the reference's electrode standard voltage relative to the SHE and is recorded as the redox potential E_H. Table 2.4 summarizes the potentials of commonly used reference electrodes relative to the SHE.

Table 2.4. Potentials (mV) of the commonly used silver-silver chloride reference electrodes versus the standard hydrogen electrode at different temperatures and different concentrations of KCl (the concentrations are related to 25 °C) (according to Galster 1991).

Temperature °C	Ag/AgCl		
	1 mol l^{-1} KCl	3 mol l^{-1} KCl	saturated KCl
0	249	224	221
5	247	221	216
10	244	217	212
15	242	214	207
20	240	211	202
25	236	207	197
30	233	203	192
35	230	200	187
40	227	196	181
45	224	192	176
50	221	188	171

Both Ag and Hg react with sulfide ions (S^{2-}) to poorly soluble AgS and HgS. Hence, direct contact between the reference electrode's diaphragm and strongly reduced environments at which SO_4^{2-} reduction takes place (Eq. 7 in Fig. 2.1) must be avoided. Otherwise, formation of solid AgS and HgS in the diaphragm will interrupt the electrical circuit. This problem may be solved by simply placing the reference electrode into a salt bridge.

2.3.1.3. Salt bridge

Redox potential values measured in the field may become erratic during the dry season because the soil moisture content is low then. Under dry conditions, only few pores remain filled with water creating a tortuous travel path for the electric current between the exposed Pt tip of the electrode and the reference electrode. Thus, the electrical circuit is not closed. A salt bridge containing potassium chloride (KCl) in which the reference electrode is set can circumvent this problem because a saturated KCl solution has a very low electrical resistance. The following instruction for preparing a salt bridge is based on the study of Veneman and Pickering (1983).

The body of the salt bridge consists of a PVC tube of a diameter somewhat larger than the reference electrode. Because a subsoil is typically wetter its topsoil, installation at depths greater than 0.5 m is usually not required. Some openings (diameter of about 5 mm) are drilled all around the tube, which are covered by a tape. These openings function as the connection reference electrode/KCl-gel/soil and enable electrical circuit during measurement. The above-ground length of the salt bridge should be 5 cm. If readings are taken periodically with a mobile voltmeter, the bridge is sealed with a cup between measurements. The salt bridge solution consists of a saturated solution of KCl (350 g l^{-1}) to which laboratory grade agar is added (about 3 wt.%). Adding phenol to prevent microbial growth within the salt bridge as described by Veneman and Pickering (1983) is not recommended. The KCl-agar-solution is heated and allowed to cool slightly to facilitate handling. The warm, viscous liquid is poured into the PVC tube, the lower end of which has been sealed with a cup, and permitted to gel.

2.3.1.4. Portable voltmeter

Ideally, no current should flow during the measurement of the redox potential (voltage) because otherwise electrochemical alterations are induced at the interface Pt surface/soil-solution due to the flow of electrons. This happens when instruments with low input resistance are used. As a result, the voltage measurement itself induces a current in the soil and the voltage measured is not representative of true soil redox conditions. Ohms law states:

$$E = I \cdot R \qquad (2.14)$$

where E is the voltage (V), I the current (A) and R the resistance (Ω). Hence, a measurement of the voltage without a current flow is impossible and theoretically, the ideal input resistance would be a value approaching infinity. Bohn (1971) outlined that instruments used for measuring redox potential in soils, which are always in the millivolt range (micro voltage), should draw negligible currents, i.e. < 1 nA. Hence, measurements of the voltage should be made with high-sensitive voltmeters with an input resistance of not less than 10^{13} Ω and a sensitivity of 1 mV. Such voltmeters are commercially available. Typically, the laboratory grade pH/E$_H$ voltmeters available in environmental laboratories fulfill this specification.

2.3.1.5. Continuous data logging

Redox conditions are not only spatially but also temporally variable, so that performing manual single time measurements at discrete intervals that are typically on the weekly, biweekly or

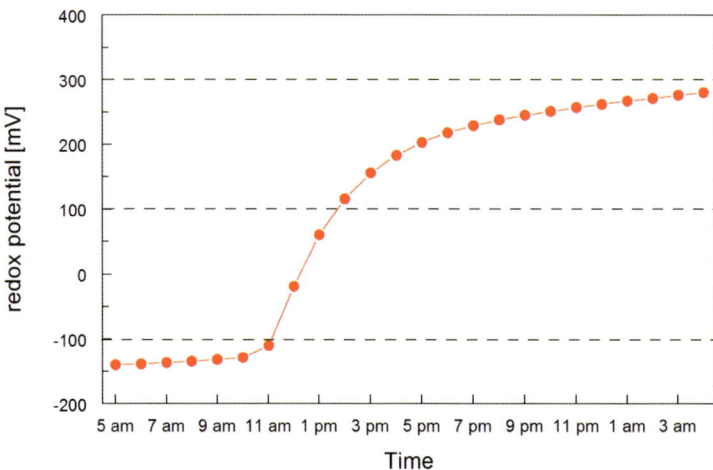

Fig. 2.4. Development of the hourly measured redox potential in the 60-cm depth of a Calcaric Gleysol during the course of 24 hours (May 17, 5:00 am to May 18, 4:00 am, 2010). A decrease in water table allowed oxygen to enter the 60-cm soil depth. The redox potentials can be classified as oxidizing (above 300 mV), weakly reducing (300 to 100 mV), moderately reducing (100 to −100 mV), and strongly reducing (below −100 mV) as indicated by the dashed lines.

monthly scale (Karathanasis et al. 2003, Mansfeldt 2003) are maybe not representative of true soil redox conditions. Figure 2.4 illustrates this matter of fact. During the course of 24 hours, the redox conditions at 60-cm depth in a marsh soil changed from strongly reducing (5:00 to 11:00 am) via moderately reducing (12:00 and 1:00 pm) to weakly reducing conditions (2:00 to 4:00 pm) in less than half a day. In case of single, manual taken measurements, the results would depend strongly on time of measurement. For example, a record at 10:00 am would classify the soil as 'strongly reducing' but at 6:00 pm as 'weakly reducing'. Hence, manual single point measurements may result in a significant loss of information in dynamic systems like soils. By installing a permanent, autonomous measuring device that measure redox potential in replication and several soil depths, this problem can be solved. Permanent installation of the Pt electrodes and automated monitoring of the redox potential have more advantages over manual measurements: remote or difficult to access, e.g., flooded study sites may be investigated; disturbance of the study site by the operator during measure by compaction of the soil and concomitant decrease in bulk density is minimized; the length of the sampling interval can be adjusted; the sensor operation can be controlled and simultaneous measurements of E_H-critical environmental parameters like pH, soil temperature, soil water content, or O_2 partial pressure are possible.

Besides Pt electrodes and a reference electrode, a control unit, a multiplexer, and a storage memory is necessary to record the readings continuously. A multiplexer increases the capacity of the data logger, allowing the connection of more electrodes for possible expansion of the system or per channel. As pointed out in section 2.3.3.1, the multiplexer and/or the logger must be equipped with an impedance for each separated redox channel greater 10^{13} Ω each in order to provide stable measurement conditions and remove any drift from the measurements. The system either can be battery- or solar-powered. If a radio network is available, transmission of the data to a web-based server can be performed by General Packet Radio Service/Internet (GPRS) (Dorau and Mansfeldt 2016).

For more details on the construction and use of continuous and autonomous data logging systems, the reader is referred to the studies of Rabenhorst (2009), Shoemaker et al. (2013), and Vorenhout et al. (2004, 2011).

2.3.2. Number of replicates

The redox potential is a variable soil parameter (Cogger et al. 1992, Norrström 1994), which accounts for local enrichment of SOM, nutrients, and microbial activity. As a result, 'hot spots' or 'micro-niches' are formed and the redox potential may significantly differ on the mm-scale (Yang et al. 2006). Flühler et al. (1976) concluded that six to nine replicates are required to obtain a representative value of the redox potential. To delineate hydric soils in problematic soil settings, at least five electrodes must be installed at a certain depth according to National Technical Committee for Hydric Soils (NTCHS 2007). Patrick et al. (1996) recommended at least three electrodes at each depth level, whereas Vepraskas and Faulkner (2001) suggested five to ten electrodes. As a rule, at least three Pt electrodes should be installed at a specific measuring depth, and budget permitting, it is desirable to install five electrodes.

2.3.3. Installation in the field

2.3.3.1. Redox electrode

A sharpened metal rod can be used to make a hole in the soil slightly larger than the diameter of the electrode body and about 2 cm less than the desired depth of the exposed Pt wire. The electrode is fed into the hole and the Pt end is pushed slightly into undisturbed soil to the measurement depth. Care should be taken that no topsoil material drops into the hole because thereby SOM, which acts as an electron donator, enters the measurement depth. If a larger set of electrodes has to be installed at one site, it is helpful to make a central point for the measuring, e.g., a small container. The Cu leads of the electrodes are led into the container. Each Cu lead is connected to a wooden lath and number-coded. By this design, it is not necessary to enter the measuring plot and disturbance of the plot is minimized (Mansfeldt 1993, Mansfeldt 2003; Fig. 2.5a). This design can also be used with a data logger system that is placed in the container. If a soil or soil

Fig. 2.5. Field monitoring stations for the measuring of the soil redox potential. The station (a) has been manually operated on a weekly basis in Lavesum, North Rhine-Westphalia, Germany. After installation, the single cables have been mounted together by cable ties. The station (b) is fully automatically operating in polder Speicherkoog at the North Sea coast of Schleswig Holstein, Germany. Besides platinum electrodes for the measuring of the redox potential at depths of 10, 20, 30, 60, 100, and 150 cm, the soil matric potential and temperature are recorded at the corres-ponding depths. Furthermore, meteorological parameters are determined. Data recording is at hourly intervals performed by a data logger system and the data are transmitted to a web-based server by General Packet Radio Service(GRPS)/Internet.

horizon is too stony, the installation of the Pt electrode is presumably impossible because the electrode body cannot be pushed into the soil at the desired depth and/or the small Pt tip will break off. Installation depths depend on the objectives of the investigations and can be specified soil depths or soil-genetically orientated, i.e. individual soil horizons. As pointed out later (section 2.3.6), the Pt electrodes should be permanently installed in the soil without replacing them. Figure 2.5b shows a study site that is equipped with Pt electrodes in different soil depths.

2.3.3.2. Salt bridge

With a standard auger, e.g., the Pürckhauer auger, a hole is made in the soil and the salt bridge is intruded after the tapes have been removed from the openings. Upon installation, the borehole is backfilled and sealed with bentonite or soil. Several Pt electrodes installed at different depths can be accommodated easily by one salt bridge. Veneman and Pickering (1983) stated, that the distance between electrode and salt bridge should remain small; they used 10 cm. However, own experiences indicates that even a distance of 10 m does not influence the readings in a marsh soil, as long as electric contact is established between the Pt and the reference electrode by water films in soil pores.

2.3.4. Measurements

For measurement, the cap of the salt bridge is removed and the reference electrode is set into the salt bridge. If the soil is very dry, the area around the salt bridge can be watered to obtain electrical contact. If no salt bridge is available, the reference electrode is pushed a short distance into the soil at the surface to ensure good electrical contact. If the soil is relatively dry, a knife is used to break up a small volume of soil and water is added to form a paste, then the reference electrode is installed in the paste to provide good contact with the soil solution. The reference electrode is connected to pH/E_H meter by means of a coaxial cable. Alligator clips and leads can be used for connecting the Pt electrodes with the pH/E_H meter. Frequently, the redox potential readings initially drift for at least several minutes before reaching a relative equilibrium or until the rate of drifting slows down considerably. The amount of drift of measured potentials is proportional to the stability of the relevant redox systems and may be attributed to O_2 desorption, Pt reduction, instability of low concentration of redox pairs, and electrode depolarization (Bohn 1971, Böttcher and Strebel 1988). In highly reduced soils and sediments, the readings stabilize quickly (within seconds), whereas in transitional systems in the process of changing from oxidizing to moderately reducing conditions (350 to -100 mV) and vice versa, the drift is significant and may persist for hours or even days. To minimize drift problems in such systems, at least an overnight stabilization period prior to making the actual measurement is suggested. If performing continuous redox potential measurements using a stabilization interface placed between the data logger and the Pt electrode, such drift can be minimized (van Bochove et al. 2002). Electrical circuit is completed by connecting the positive lead of the voltmeter to a Pt electrode and the negative lead to the reference electrode.

2.3.5. Interpreting the results of redox potential measurements

2.3.5.1. Calculation of redox potentials and related parameters

When the measuring system has equilibrated, the voltage indicated by the voltmeter is recorded. The redox potential is calculated from the voltage by the following equation:

$$E_H = E_{ref} + f \qquad (2.15)$$

where E_H is the redox potential against the SHE (mV), E_{ref} the measured value against the reference electrode (mV), and f is a factor (mV) considering the kind of the reference electrode (Table 2.4).

For example, if soil's pH is 6 and if the reading is 50 mV and an Ag/AgCl (3 M KCl) reference electrode was used at a soil temperature of 15 °C, f is 214 mV. As a result, the redox potential is 264 mV. The corresponding pe value is 4, and pe+pH is 10. As redox potential is most often a mixed potential, measurement more precise than ±10 mV has little significance (Mueller et al. 1985). Hence, the calculated redox potential in this example should be reported as 260 mV. Because the temperature-dependent correction factor for the reference electrode used is rather small (Table 2.4), corrections for field temperature can be omitted and a factor at constant temperature, e.g., 25 °C, can be applied (Patrick et al. 1996). Typically, redox potentials of soils range from 800 mV (oxidizing) to –400 mV (strongly reducing).

2.3.5.2. Interpretation of redox potentials

Knowing the present redox status of soils is crucial, especially for Fe but also for Mn compounds, because both elements are (in contrast to aluminum, Al) redox-active, and hence, participate in redox reactions. In their oxidized species, Fe and Mn occur as strongly colored oxides in soils, whereas under reducing conditions these oxides are reductively dissolved, liberating their water-soluble reduced counterparts, ferrous Fe (Fe^{2+}) and manganous Mn (Mn^{2+}) (Eq. 3 and (Eq. 4) in Fig. 2.1).

Typically, repeated changes in oxidation and reduction are reflected in special patterns of soil color, which are called redoximorphic features. The associated process is called as *redoximorphosis*. The occurrence of redoximorphic features is used in many national and international soil classification systems to infer reducing conditions in soils (e.g., Ad-hoc AG Boden 2005, IUSS Working Group WRB 2015, Soil Survey Staff 2014). However, because soil color may not always reflect the current redox conditions of a soil, for instance due to the intrinsic color of the soil's parent material, high content of organic matter, or where soil color preserves relict redox conditions, methods of recording reducing conditions in soils are necessary for understanding soil formation and performing soil classification.

Furthermore, redox conditions play a key role in the behavior and fate of many elements and compounds occurring as either nutrients or pollutants in soils.

- Iron oxides are strong adsorbents of toxic (semi)metals such as arsenic (Bowell 1994) or nutrients such as phosphate (Reddy et al. 1999). As these Fe oxides are dissolved by reduction, the compounds adsorbed on them may be released (Mansfeldt and Overesch 2013, Peretyazhko and Sposito 2005), making them mobile and bioavailable.
- The same process, namely reductive dissolution, dissolves Mn oxides and adsorbed metals such as molybdenum (Mo) and cobalt (Co), which have a strong affinity for Mn oxides, can be liberated (Hindersmann and Mansfeldt 2014). Note that this process begins under weakly reducing conditions.
- Both nitrous oxide (N_2O) and methane (CH_4) are potent greenhouse gases. Nitrous oxide is produced in soils under weakly reducing conditions and methane under strongly reducing conditions (Yu and Patrick 2003, Yu and Patrick 2004).
- Under strongly reducing conditions, sulfate is reduced to sulfide. Simultaneously, trace elements may be precipitated as poorly soluble metal sulfides such as arsenic (As) sulfides (Onstott et al. 2011) or copper (Cu) sulfides (Weber et al. 2009). Precipitation lowers the solubility of the metals.
- Some trace elements are directly involved in electron transfer and, hence, occur in different valences. For example, antimony (Sb) with its oxidized species antimonate (e.g., $Sb^V(OH)_6^-$)

and reduced species antimonite (e.g., $Sb^{III}(OH)_3^0$). Redox-dependent speciation is important, because oxidized and reduced species often have different toxicity and solubility (Smedley and Kinniburgh 2002). For example, a change in redox conditions from oxidizing to weakly reducing in soil suspension resulted in a decrease of the concentration of dissolved Sb, which was thought to be a result of the reduction of the soluble antimonate to less soluble antimonite (Hindersmann and Mansfeldt 2014).

- Sorption and biotransformation of many organic pollutants in soils, such as pharmaceuticals or biocides, are controlled by redox state (Crawford et al. 2000, Dalkmann et al. 2014, Mohatt et al. 2011).
- Relationships were found between redox potential and the distribution of certain plant species (Josselyn et al. 1990, Pennington and Walters 2006).
- Bacterial communities may respond to changes in redox potential along a moisture gradient in restored wetlands (Peralta et al. 2014).

2.3.5.3. Restrictions in the interpretation of redox potentials

Proper interpretation of redox potential measurements is difficult from both a theoretical and practical point of view.

In general, thermodynamic limitations restrict the interpretations of measured redox potential (Bohn 1971, Whitfield 1974). The ability of the Nernst equation to predict the activities of redox species quantitatively is, strictly speaking, only valid under chemical equilibrium conditions. However, due to the permanent input of organic matter (electron donor) and its microbial-driven decomposition, there is a continual release and uptake of electrons in soils. Hence, redox conditions in soils are metastable and the dynamic disequilibrium state resulting from this metastability is typical for soils.

For a Pt electrode to respond to a specific redox pair, the pair must be electroactive, i.e. electron transfer reactions must be rapid and reversible in order to attain equilibrium. Furthermore, both members of the pair must be present in the soil solution at concentrations greater than about 10^{-5} M. However, many of the important redox reactions involving the elements C, N, and S are irreversible because non-electroactive gases and molecules (e.g., N_2, CH_4; see Fig. 2.1) are consumed or formed. In contrast, Mn and Fe are forming electroactive redox couples, which generate sufficient anodic and cathodic currents to obtain a measurable voltage of the system. Hence, the Pt electrode works well when the range of Mn and Fe reduction is achieved in soils. Under oxidizing conditions, however, the extremely low solubility of the oxidized species of these metals ($Mn^{3+/4+}$, Fe^{3+}) results in activities far below 10^{-5} M, which generates anodic and cathodic currents too low for obtaining a measurable voltage.

Electrode contamination (sometimes called 'electrode poisoning') can be defined as any chemical or physical change in the Pt electrode that prevents or hinders superficial electron exchange (Devitt et al. 1989). Practically, electrode contamination refers to the development of coatings and precipitates on the Pt surface, which are composed of salts, carbonates, organic carbon, sulfide, but also O_2. Adsorbed species also affect the measured potential. Under oxidizing conditions, O_2 is adsorbed onto the Pt surface where it forms $Pt(OH)_2$ which gives rise to a potential of +568 mV at pH 7 (Eq. 2.16). This potential is unreliable as a measure of dissolved O_2 status and hence Pt electrode potentials can be used only as empirical values for comparison purposes and to draw a conclusion that 'oxidizing conditions' are prevailing. Furthermore, Eq. 2.16 reveals that the oxidized Pt surface acts as an oxide electrode that responds to the activity of H^+, i.e. pH, rather than to O_2 partial pressure:

$$Pt(OH)_2 + 2e^- + 2H^+ = Pt^0 + 2H_2O \quad (E^0{}_7 = 568 \text{ mV}) \quad (2.16)$$

Under strongly reducing conditions and in the presence of sulfate, its reduced counterpart sulfide reacts with the Pt surface to form black platinum sulfide (PtS, Eq. 2.17):

$$PtS + 2e^- + 2H^+ = Pt^0 + H_2S \ (E^0_7 = -710 \text{ mV}) \quad (2.17)$$

Platinum sulfide coats are extremely stable, determine the electrode potentials according to Eq. 2.17 and render the Pt electrode insensitive. Particularly in estuarine or marine sediments, rich in sulfate, these coatings prevent a long-term field installation of the Pt electrode.

Table 2.5 demonstrates the idealized sequential redox sequence observed in soils and the individual redox potential at pH 7 assuming realistic activities for gases and ions. Additionally, empirically determined redox potentials measured with Pt electrodes are listed. Clearly, theoretical and empirical potentials differ significantly, which can be explained with the limitations outlined above (reaction kinetics, irreversibility, non-equilibrium, etc.). Overall, true Nernstian behavior cannot be assumed, i.e. the redox species activities are far from any quantitative interpretation, and in the best case, redox potentials may be interpreted in a semi-quantitative sense. Therefore, the use of redox classes or zones as presented in Fig. 2.1 over the use of numerical values is encouraged.

Table 2.5. Order of utilization of principal electron acceptors in soils, reduction potentials of these half-reactions at pH 7 (E_{H7}), and measured redox potentials (E_H) of these reactions in soils (according to Strawn et al. 2015). [a] Activities for soluble ions and trace gases = 10^{-4}, O_2 = 0.21, N_2 = 0.78, CO_2 = 0.00032, solids = 1.

Reaction	E_{H7} (mV)[a]	Measured E_H in soils (mV)
O_2 disappearance $O_2 + 4e^- + 4H^+ = H_2O$	800	600 to 400
NO_3^- disappearance via denitrification $2NO_3^- + 10e^- + 12H^+ = N_2 + 6H_2O$	700	500 to 200
Mn^{2+} formation $MnO_2 + 2e^- + 4H^+ = Mn^{2+} + 2H_2O$	520	400 to 200
Fe^{2+} formation from amorphous Fe oxides $Fe(OH)_3 + e^- + 3H^+ = Fe^{2+} + 2H_2O$	−71	300 to 100
H_2S formation $SO_4^- + 5e^- + 10H^+ = H_2S + 4H_2O$	−210	0 to −150
Fe^{2+} formation from goethite $FeOOH + e^- + 3H^+ = Fe^{2+} + 2H_2O$	−230	300 to 100
CH_4 formation $CO_2 + 8e^- + 8H^+ = CH_4 + 2H_2O$	−240	−100 to −200
H_2 formation $2H^+ + 2e^- = H_2$	−410	−150 to −220

2.3.6. Maintenance

2.3.6.1. Reliability, checking and cleaning of the platinum electrode

Measurements taken immediately (within a few minutes to a few hours) after redox electrode installation, i.e. temporary installation, may be skewed by installation disturbance, introduction of O_2 to a reducing environment and drift in the readings (see section 2.3.4). Thomas et al. (2009)

found that readings from permanent electrodes had much lower variability and were more useful for ecological comparisons than those from temporary electrodes were. Hence, it is emphatically recommended to avoid such short installation time. Instead, Pt electrodes should be placed at least for several weeks in the soil but there is nothing wrong to extend this period, i.e. permanent installation.

There is no broad agreement how long Pt electrodes may remain in the field before malfunctioning. On the one hand, both Rickman et al. (1968) and Devitt et al. (1989) suggested that any adverse effect due to Pt contamination (precipitation of calcium carbonate, alumosilicates) could be minimized by removing the electrodes after 2 months of continuous operation in soils. D'Amore et al. (2015) replaced and reconditioned redox electrodes on a regular basis and Niedermeier and Robinson (2007) limited measurements of redox potential to one year partly to minimize the risk of Pt electrode failure. Reddy and DeLaune (2008) recommended checking electrodes periodically (at least once every three months). On the other hand, no significant electrode failure without replacement of electrodes has been reported after installing electrodes for a few months (Jordan et al. 1993, Olness et al. 1989, Teichert et al. 2000), one year (Comerford et al. 1996, Josselyn et al. 1990, Vepraskas and Wilding 1983), two years (Johnston et al. 1995, Megonigal et al. 1993), three years (Armstrong et al. 1985, Faulkner and Patrick 1992), four years (Mansfeldt 2003, Thompson et al. 1998), five years (Austin and Huddleston 1999, Reuter and Bell 2001), to up to nine years (Jenkinson et al. 2002) in soils. When evaluating these discrepancies, it should be noted that Rickman et al. (1968) and Devitt et al. (1989) determined the oxygen diffusion rate (ODR) in soils in which a current is applied to the Pt electrode. Obviously, electrochemical alterations have been induced by this current at the Pt/soil-solution interface due to flow of electrons. In contrary, the current occurring during the determination of the redox voltage is negligible (see section 2.3.1.4) and formation of coatings (with the exception of oxide coatings) can rather be disregarded.

To examine field-installed Pt electrodes they must be removed, rinsed with demineralized water and then immersed into a diluted redox buffer solution. The use of undiluted redox buffers should be avoided, because these solutions are highly poised, which means that they are insensitive to differences in the response of the electrodes, which are significant when measurements in soils, which are usually of low poise, are made. Because undiluted redox buffer solutions produce a stable potential they are not capable of resolving between small differences in Pt electrode performance, which may be caused by adsorption of compounds onto the surface of the electrode. Mansfeldt (1993) used demineralized water to check electrodes but this system is very poorly poised, and therefore, not recommended. Instead, a 1:100 Zobell's solution is more suitable (Teasdale et al. 1998).

Malfunctioning electrodes can be identified by strongly oscillating readings (in the range of seconds to minutes) or 'unrealistic' values. Whether redox data are unrealistic, e.g., no electrode response during a drying-rewetting cycle, or not is not easy to decide because natural redox potentials display strong variations. The higher the number of replicates and the more other E_H-sensitive parameters are known, e.g., water table or soil temperature, the easier the decision is. If electrode failure is suspected, the electrode must be removed and cleaned. The most effective way to clean Pt electrodes is to use abrasive polishing by 0.005 mm alumina slurry on a felt pad, to dip the electrode into a 10% (v/v) nitric acid solution, to rinse the electrode with demineralized water and then to verify the electrode in a diluted redox buffer. Contaminated electrodes should function properly again after this cleaning procedure. If this fails to help, the electrode is broken and must be removed and repaired.

Replacing electrodes, reconditioning and reinstalling them on a regular basis has been done by some researchers independent of season (D'Amore et al. 2015) or during dry summers when redox potentials were oxidizing (He et al. 2003, Vepraskas et al. 2004). No general advice can be given if this approach is correct or not. Even if electrodes are removed and later reinstalled in exactly the same site, there is no guarantee that the replaced electrodes 'see' the same soil microenvironment ('micro niches') as before.

To summarize: if properly constructed, especially when fitted with the ceramic jacket, Pt electrodes may be permanently installed in soils even in permanently waterlogged settings. The only conditions unsuitable for permanent electrode deployment are strongly reducing environments where intensive sulfate reduction takes place, e.g., marine environments. Applying a current to the Pt electrode in order to measure ODR must be avoided.

2.3.6.2. Reference electrode

Farrell et al. (1991) stated that commercial reference electrodes are generally unsuitable for in-situ monitoring programs, because they are relatively expensive, not very rugged, and have only a small salt bridge reservoir that requires frequent refilling. Nowadays, reference electrodes are available whose electrolyte is gel-stabilized (Maksymiuk et al. 2013). This minimizes leaching of the electrolyte so that these electrodes are suitable for long-term measurement of the redox potential, especially when they are placed into a salt bridge. However, gel aging and the impossibility of refilling the electrolyte are some drawbacks but in the meantime, electrodes are available whose electrolyte reservoir can be replenished. The author own experiments have revealed that such electrodes function properly in the field consecutively for time periods of at least two years.

2.3.6.3. Salt bridge

Malfunction of the salt bridge can be caused by compression of the gel a longer time after installation (Mansfeldt 1993). For this reason, the gel should be replaced from time to time. This can be done without exchanging the salt bridge body. Niedermeier and Robinson (2007), for example, replaced their salt bridge every three months.

2.4. References

Ad-hoc AG Boden, 2005. Bodenkundliche Kartieranleitung. 5. Auflage. Schweizerbart Science Publishers, Stuttgart, Germany.

Armstrong, W., Wright, E.J., Lythe, S., Gaynard, T.J., 1985. Plant zonation and the effects of the spring-neap tidal cycle on soil aeration in a Humber salt marsh. J. Ecol. 73: 323–339.

Austin, W.E., Huddleston, J.H., 1999. Viability of permanently installed platinum redox electrodes. Soil Sci. Soc. Am. J. 63: 1757–1762.

Biddle, D.L., Chittleborough, D.J., Fitzpatrick, R.W., 1995. Field-based comparison of platinum and wax impregnated graphite redox electrodes. Aust. J. Soil Res. 33: 415–424.

Bohn, H.L., 1971. Redox potentials. Soil Sci. 112: 39–45.

Bohn, H.L., McNeal, B.L., O'Connor, G.A., 2001. Soil chemistry. 3rd ed. John Wiley & Sons, New York, USA.

Böttcher, J., Strebel, O., 1988. Ermittlung des Redoxpotentials (Eh-Wert) von Böden und Grundwässern aus dem zeitlichen Verlauf der Elektrodenpolarisation. Z. Pflanzenernähr. Bodenk. 151: 363–368.

Bowell, R.J., 1994. Sorption of arsenic by iron-oxides and oxyhydroxides in soils. Appl. Geochem. 9: 279–286.

Castenson, K.L., Rabenhorst, M.C., 2006. Indicator of reduction in soil (IRIS): Evaluation of a new approach for assessing reduced conditions in soil. Soil Sci. Soc. Am. J. 70: 1222–1226.

Chapelle, F.H., McMahon, P.B., Dubrovsky, N.M., Fujii, R.F., Oaksford, E.T., Vroblesky, D.A., 1995. Deducing the distribution of terminal electron-accepting process in hydrologically diverse groundwater systems. Water Resour. Res. 31: 359–371.

Childs, C.W., 1981. Field tests for ferrous iron and ferric-organic complexes (on exchange sites or in wa-

ter-soluble forms) in soils. Aust. J. Soil Res. 19: 175–180.
Cogger, C.G., Kennedy, P.E., Carlson, D., 1992. Seasonally saturated soils in the Puget lowland II. Measuring and interpreting redox potentials. Soil Sci. 154: 50–58.
Comerford, N.B., Jerez, A., Freitas, A.A., Montgomery, J., 1996. Soil water table, reducing conditions, and hydrologic regime in a Florida flatwood landscape. Soil Sci. 161: 194–199.
Crawford, J.J., Traina, S.J., Tuovinen, O.H., 2000. Bacterial degradation of atrazine in redox potential gradients in fixed-film sand columns. Soil Sci. Soc. Am. J. 64: 624–634.
Dalkmann, P., Dresemann, T.F., Siebe, C., Mansfeldt, T., Amelung, W., Siemens, J., 2014. Release of pharmaceuticals under reducing conditions in a wastewater-irrigated Mexican soil. J. Environ. Qual. 43: 1926–1932.
D'Amore, D.V., Ping, C.L., Herendeen, P.A., 2015. Hydromorphic soil development in the coastal temperate rainforest of Alaska. Soil Sci. Soc. Am. J. 79: 698–709.
Devitt, D.A., Stolzy, L.H., Miller, W.W., Campana, J.E., Sternberg, P., 1989. Influence of salinity, leaching fraction, and soil type on oxygen diffusion rate measurements and electrode "poisoning". Soil Sci. 148: 327–335.
Dirasian, H.A., 1968. Electrode potentials-significance in biological systems. Water Sewage Works 115: 420–425.
Dorau, K., Eickmeier, M., Mansfeldt, T., 2016. Comparison of manganese and iron oxide-coated redox bars for characterization of the redox status in wetland soils. Wetlands 36: 133–141.
Dorau, K., Mansfeldt, T., 2015. Manganese-oxide-coated redox bars as an indicator of reducing conditions in soils. J. Environ. Qual. 44: 696–703.
Dorau, K., Mansfeldt, T., 2016. Comparison of redox potential dynamics in a diked marsh soil: 1990 to 1993 versus 2011 to 2014. J. Plant Nutr. Soil Sci. 179: 641–651.
Dusek, J., Picek, T., Cizkova, H., 2008. Redox potential dynamics in a horizontal subsurface flow constructed wetland for wastewater treatment: Diel, seasonal and spatial fluctuations. Ecol. Eng. 34: 223–232.
Essington, M.E., 2015. Soil and water chemistry: An integrative approach. 2nd ed. CRC Press, Boca Raton, FL, USA.
Fakih, M., Davranche, M., Dia, A., Nowack, B., Petitjean, P., Chatellier, X., Gruau, G., 2008. A new tool for in situ monitoring of Fe-mobilization in soils. Appl. Geochem. 23: 3372–3383.
Farrell, R.E., Swerhone, G.D.W., Vankessel, C., 1991. Construction and evaluation of a reference electrode assembly for use in monitoring in situ soil redox potentials. Commun. Soil Sci. Plant Anal. 22: 1059–1068.
Faulkner, S.P., Patrick, W.H.J., 1992. Redox processes and diagnostic wetland soil indicators in bottomland hardwood forest. Soil Sci. Soc. Am. J. 56: 856–865.
Flühler, H., Ardakani, M.S., Szuszkiewicz, T.E., Stolzy, L.H., 1976. Field-measured nitrous oxide concentrations, redox potentials, oxygen diffusion rates, and oxygen partial pressures in relation to denitrification. Soil Sci. 122: 107–114.
Galster, H., 1991. pH measurement: Fundamentels, methods, applications, instrumentation. VCH Verlagsgesellschaft, Weinheim, Germany.
Gao, S., Tanji, K.K., Scardaci, S.C., Chow, A.T., 2002. Comparison of redox indicators in a paddy soil during rice-growing season. Soil Sci. Soc. Am. J. 66: 805–817.
Gerwing, T.G., Gerwing, A.M.A., Drolet, D., Hamilton, D.J., Barbeau, M.A., 2013. Comparison of two methods of measuring the depth of the redox potential discontinuity in intertidal mudflat sediments. Mar. Ecol.-Prog. Ser. 487: 7–13.
Gillespie, L.J., 1920. Reduction potentials of bacterial cultures and of water-logged soils. Soil Sci. 9: 199–216.
He, X., Vepraskas, M.J., Lindbo, D.L., Skaggs, R.W., 2003. A method to predict soil saturation frequency and duration from soil color. Soil Sci. Soc. Am. J. 67: 961–969.
Heduit, A., Martin, B., Duchamp, I., Thevenot, D.R., 1993. Comparison of gold and platinum-electrode responses in activated-sludge. Water Sci. Technol. 28: 473–480.
Hinchey, E.K., Schaffner, L.C., 2005. An evaluation of electrode insertion techniques for measurement of redox potential in estuarine sediments. Chemosphere 59: 703–710.
Hindersmann, I., Mansfeldt, T., 2014. Trace element solubility in a multimetal-contaminated soil as affected by redox conditions. Water Air Soil Pollut. 225: 2158.
IUSS Working Group WRB., 2015. World Reference Base for Soil Resources 2014, update 2015. World

Soil Resources Reports No. 106. FAO, Rome, Italy.

Jang, A., Lee, J.H., Bhadri, P.R., Kumar, S.A., Timmons, W., Beyette, F.R., Papautsky, I., Bishop, P.L., 2005. Miniaturized redox potential probe for in situ environmental monitoring. Environ. Sci. Technol. 39: 6191–6197.

Jenkinson, B.J., Franzmeier, D.P., 2006. Development and evaluation of iron-coated tubes that indicate reduction in soils. Soil Sci. Soc. Am. J. 70: 183–191.

Jenkinson, B.J., Franzmeier, D.P., Lynn, W.C., 2002. Soil hydrology on an end moraine and a dissected till plain in west-central Indiana. Soil Sci. Soc. Am. J. 66: 1367–1376.

Johnston, C.A., Pinay, G., Arens, C., Naiman, R.J., 1995. Influence of soil properties on the biogeochemistry of a beaver meadow hydrosequence. Soil Sci. Soc. Am. J. 59: 1789–1799.

Jordan, T.E., Correll, D.L., Weller, D.E., 1993. Nutrient interception by a riparian forest receiving inputs from adjacent cropland. J. Environ. Qual. 22: 467–473.

Josselyn, M.N., Faulkner, S.P., Patrick, W.H.J., 1990. Relationships between seasonally wet soils and occurrence of wetland plants in California. Wetlands 10: 7–26.

Karathanasis, A.D., Thompson, Y.L., Barton, C.D., 2003. Long-term evaluations of seasonally saturated "wetlands" in western Kentucky. Soil Sci. Soc. Am. J. 67: 662–673.

Kleimeier, C., Karsten, U., Lennartz, B., 2014. Suitability of degraded peat for constructed wetlands – Hydraulic properties and nutrient flushing. Geoderma 228: 25–32.

Light, T.S., 1972. Standard solution for redox potential measurements. Anal. Chem. 44: 1038–1039.

Lindsay, W.L., 1979. Chemical equilibria in soils. John Wiley & Sons, New York, USA.

Maksymiuk, K., Michalska, A., Kisiel, A., Galus, Z., 2013. Silver electrodes. In: G. Inzelt, A. Lewenstam and F. Scholz, editors, Handbook of Reference Electrodes. Springer, Berlin Heidelberg, Germany. p. 86–105.

Mansfeldt, T., 1993. Redoxpotentialmessungen mit dauerhaft installierten Platinelektroden unter reduzierenden Bedingungen. Z. Pflanzenernähr. Bodenk. 156: 287–292.

Mansfeldt, T., 2003. In situ long-term redox potential measurements in a dyked marsh soil. J. Plant Nutr. Soil Sci. 166: 210–219.

Mansfeldt, T., Overesch, M., 2013. Arsenic mobility and speciation in a Gleysol with petrogleyic properties: A field and laboratory approach. J. Environ. Qual. 42: 1130–1141.

Megonigal, J.P., Patrick, W.H.J., Faulkner, S.P., 1993. Wetland identification in seasonally flooded forest soils: soil morphology and redox dynamics. Soil Sci. Soc. Am. J. 57: 140–149.

Mohatt, J.L., Hu, L.H., Finneran, K.T., Strathmann, T.J., 2011. Microbially mediated abiotic transformation of the antimicrobial agent sulfamethoxazole under iron-reducing soil conditions. Environ. Sci. Technol. 45: 4793–4801.

Mueller, S.C., Stolzy, L.H., Fick, G.W., 1985. Constructing and screeening platinum microelectrodes for measuring soil redox potential. Soil Sci. 139: 558–560.

Niedermeier, A., Robinson, J.S., 2007. Hydrological controls on soil redox dynamics in a peat-based, restored wetland. Geoderma 137: 318–326.

Norrström, A.C., 1994. Field-measured redox potentials in soils at the groundwater surface-water interface. Eur. J. Soil Sci. 45: 31–36.

NTCHS, National Technical Committee for Hydric Soils., 2007. The hydric soil technical standard. Technical Note 11. National Technical Committee for Hydric Soils. p. 31.

Olness, A., Rinke, J., Hung, H.-M., Evans, S.D., 1989. Effect of tillage on redox potential of a tara silt loam soil. Soil Sci. 148: 265–274.

Olsen, K.R., Andersen, F.O., 1994. Nutrient cycling in shallow, oligotrophic lake Kvie, Denmark. Hydrobiologia 275: 255–265.

Onstott, T.C., Chan, E., Polizzotto, M.L., Lanzon, J., DeFlaun, M.F., 2011. Precipitation of arsenic under sulfate reducing conditions and subsequent leaching under aerobic conditions. Appl. Geochem. 26: 269–285.

Ottow, J.C.G., 2011. Mikrobiologie von Böden: Biodiversität, Ökophysiologie und Metagenomik. Springer, Berlin Heidelberg, Germany.

Owens, P.R., Wilding, L.P., Lee, L.M., Herbert, B.E., 2005. Evaluation of platinum electrodes and three electrode potential standards to determine electrode quality. Soil Sci. Soc. Am. J. 69: 1541–1550.

Owens, P.R., Wilding, L.P., Miller, W.M., Griffin, R.W., 2008. Using iron metal rods to infer oxygen status in seasonally saturated soils. Catena 73: 197–203.

Pang, H., Zang, T., 1998. Fabrication of redox potential microelectrodes for studies in vegetated soils of biofilm systems. Environ. Sci. Technol. 32: 3646–3652.
Patrick, W.H., Gambrell, R.P., Faulkner, S.P., 1996. Redox measurement of soils. In: Sparks, D.L. (ed.), Methods of soil analysis. 3rd ed. Soil Science Society of America, Madison, WI, USA. p. 1225–1273.
Patrick, W.H.J., 1966. Apparatus for controlling the oxidation-reduction potential of waterlogged soils. Nature 212: 1278–1279.
Pennington, M.R., Walters, M.B., 2006. The response of planted trees to vegetation zonation and soil redox potential in created wetlands. For. Ecol. Manage. 233: 1–10.
Peralta, A.L., Ludmer, S., Matthews, J.W., Kent, A.D., 2014. Bacterial community response to changes in soil redox potential along a moisture gradient in restored wetlands. Ecol. Eng. 73: 246–253.
Peretyazhko, T., Sposito, G., 2005. Iron(III) reduction and phosphorous solubilization in humid tropical forest soils. Geochim. Cosmochim. Ac. 69: 3643–3652.
Pfisterer, U., Gribbohm, S., 1989. Zur Herstellung von Platinelektroden für Redoxmessungen. Z. Pflanzenernähr. Bodenk. 152: 455–456.
Plante, R., Alcalado, P.M., Martineziglesias, J.C., Ibarzabal, D., 1989. Redox potential in water and sediments of the Gulf of Batabano, Cuba. Estuar. Coast. Shelf Sci. 28: 173–184.
Ponnamperuma, F.N., 1972. The chemistry of submerged soils. Adv. Agron. 24: 29–96.
Rabenhorst, M.C., 2009. Making soil oxidation–reduction potential measurements using multimeters. Soil Sci. Soc. Am. J. 73: 2198–2201.
Rabenhorst, M.C., Burch, S.N., 2006. Synthetic iron oxides as an indicator of reduction in soils (IRIS). Soil Sci. Soc. Am. J. 70: 1227–1236.
Rabenhorst, M.C., Hively, W.D., James, B.R., 2009. Measurements of soil redox potential. Soil Sci. Soc. Am. J. 73: 668–674.
Reddy, K.R., DeLaune, R.D., 2008. Biogeochemistry of wetlands: Science and applications. CRC Press, Boca Raton, FL, USA.
Reddy, K.R., Kadlec, R.H., Flaig, E., Gale, P.M., 1999. Phosphorus retention in streams and wetlands: A review. Crit. Rev. Environ. Sci. Technol. 29: 83–146.
Rennert, T., Mansfeldt, T., 2005. Iron-cyanide complexes in soil under varying redox conditions: speciation, solubility and modelling. Eur. J. Soil Sci. 56: 527–536.
Reuter, R.J., Bell, J.C., 2001. Soils and hydrology of a wet-sandy catena in east-central Minnesota. Soil Sci. Soc. Am. J. 65: 1559–1569.
Rezanezhad, F., Couture, R.M., Kovac, R., O'Connell, D., Van Cappellen, P., 2014. Water table fluctuations and soil biogeochemistry: An experimental approach using an automated soil column system. J. Hydrol. 509: 245–256.
Rickman, R.W., Letey, J., Aubertin, G.M., Stolzy, L.H., 1968. Platinum microelectrode poisoning factors. Soil Sci. Soc. Am. Proc. 32: 204–208.
Scott, M.J., Morgan, J.J., 1990. Energetics and conservative properties of redox systems. In: D. C. Melchior and R. L. Barrett, editors, Chemical modeling of aqueous systems II. American Chemical Society, Washington DC. p. 368–378.
Shoemaker, C., Kroger, R., Reese, B., Pierce, S.C., 2013. Continuous, short-interval redox data loggers: verification and setup considerations. Environ. Sci.: Process Impacts 15: 1685–1691.
Smedley, P.L., Kinniburgh, D.G., 2002. A review of the source, behaviour and distribution of arsenic in natural waters. Appl. Geochem. 17: 517–568.
Soil Survey Staff 2014. Keys to soil taxonomy. 12th ed. USDA-Natural Resources Conservation Service, Washington DC, USA.
Stiles, C.A., Dunkinson, E.T., Ping, C.L., Kidd, J., 2010. Initial field installation of manganese indicators of reduction in soils, Brooks Range, Alaska. Soil Survey Horizon 51: 102–107.
Strawn, D.G., Bohn, H.L., O'Connor, G.A., 2015. Soil chemistry. 4th ed. Wiley Blackwell, Chichester, UK.
Swerhone, G.D.W., Lawrence, J.R., Richards, J.G., Hendry, M.J., 1999. Construction and testing of a durable platinum wire Eh electrode for in situ redox measurements in the subsurface. Ground Water Monit. Remediat. 19: 132–136.
Teasdale, P.R., Minett, A.I., Dixon, K., Lewis, T.W., Batley, G.E., 1998. Practical improvements for redox potential (EH) measurements and the application of a multiple-electrode redox probe (MERP) for characterising sediment in situ. Anal. Chim. Acta 367: 201–213.
Teichert, A., Böttcher, J., Duijnisveld, W.H.M., 2000. Redox measurements as a qualitative indicator of

spatial and temporal variability of redox state in a sandy forest soil. In: Schüring, J., Schulz, H.D., Fischer, W.R., Böttcher, J., Duijnisveld, W.H.M., (eds.), Redox: fundamentals, processes and applications. Springer, Berlin, Germany. p. 95–110.

Thomas, C.R., Miao, S.L., Sindhoj, E., 2009. Environmental factors affecting temporal and spatial patterns of soil redox potential in Florida Everglades wetlands. Wetlands 29: 1133–1145.

Thompson, J.A., Bell, J.C., ZannerC.W., 1998. Hydrology and hydric soil extent within a mollisol catena in Southeastern Minnesota. Soil Sci. Soc. Am. J. 62: 1126–1133.

van Bochove, E., Beauchemin, S., Theriault, G., 2002. Continuous multiple measurement of soil redox potential using platinum microelectrodes. Soil Sci.Soc. Am. J. 66: 1813–1820.

Veneman, P.L.M., Pickering, E.W., 1983. Salt bridge for field redox potential measurements. Commun. Soil Sci. Plant Anal. 14: 669–677.

Vepraskas, M.J., Faulkner, S.P., 2001. Redox chemistry of hydric soils. In: Richardson, J.L., Vepraskas, M.J. (eds.), Wetland Soils: Genesis, hydrology, landscapes, and classification. CRC Press, Boca Raton, FL, USA. p. 85–105.

Vepraskas, M.J., He, X., Lindbo, D.L., Skaggs, R.W., 2004. Calibrating hydric soil field indicators to long-term wetland hydrology. Soil Sci. Soc. Am. J. 68: 1461–1469.

Vepraskas, M.J., Wilding, L.P., 1983. Aquic moisture regimes in soils with and without low chroma colors. Soil. Sci. Soc. Am. J. 47: 280–285.

Vorenhout, M., van der Geest, H.G., Hunting, E.R., 2011. An improved datalogger and novel probes for continuous redox measurements in wetlands. Int. J. Environ. Anal. Chem. 91: 801–810.

Vorenhout, M., van der Geest, H.G., van Marum, D., Wattel, K., Eijsackers, H.J.P., 2004. Automated and continuous redox potential measurements in soil. J. Environ. Quality 33: 1562–1567.

Wafer, C.C., Richards, J.B., Osmond, D.L., 2004. Construction of platinum-tipped redox probes for determining soil redox potential. J. Environ. Qual. 33: 2375–2379.

Weber, F.A., Voegelin, A., Kretzschmar, R., 2009. Multi-metal contaminant dynamics in temporarily flooded soil under sulfate limitation. Geochim. Cosmochim. Ac. 73: 5513–5527.

Weigand, H., Mansfeldt, T., Bäumler, R., Schneckenburger, D., Wessel-Bothe, S., Marb, C., 2010. Arsenic release and speciation in a degraded fen as affected by soil redox potential at varied moisture regime. Geoderma 159: 371–378.

Whisler, F.D., Lance, J.C., Linebarger, R.S., 1974. Redox potentials in soil columns intermittently flooded with sewage water. J. Environ. Qual. 3: 68–74.

Whitfield, M., 1974. Thermodynamic limitations on the use of the platinum electrode in Eh measurements. Limnol. Oceanogr. 19: 857–865.

Yang, J., Hu, Y.M., Bu, R.C., 2006. Microscale spatial variability of redox potential in surface soil. Soil Sci. 171: 747–753.

Yu, K.W., Patrick, W.H., 2003. Redox range with minimum nitrous oxide and methane production in a rice soil under different pH. Soil Sci. Soc. Am. J. 67: 1952–1958.

Yu, K.W., Patrick, W.H., 2004. Redox window with minimum global warming potential contribution from rice soils. Soil Sci. Soc. Am. J. 68: 2086–2091.

Zehnder, A.J.B., Wuhrmann, K., 1976. Titanium(III) citrate as a nontoxic oxidation-reduction buffering system for culture of obligate anaerobes. Science 194: 1165–1166.

Zhi-Guang, L., 1985. Oxidation-reduction potential. In: Y. Tian-ren, editor Physical chemistry of paddy soils. Springer, Berlin, Germany. p. 1–26.

ZoBell, C.E., 1946. Studies on redox potential of marine sediments. Bull. Am. Assoc. Petrol. Geol. 30: 477–513.

3. Soil pH value

S. Thiele-Bruhn

3.1. Introduction

3.1.1. Background on pH

The pH value or 'soil reaction' is a key parameter governing the vast majority of soil chemical and physicochemical as well as biological processes and functions. These include the surface charge of soil colloids such as clay minerals and humic substances and the resulting aggregation, as well as the retardation or mobility of nutrients and polar chemicals, the speciation of metals and organic molecules, organic matter complexation and release (Blume et al. 2011, Bolan et al. 2004, Grybos et al. 2007, Tan 2011). The growth of plants and the abundance and biodiversity of soil biota as well as their functions and activity, e.g. in biogeochemical cycling of nutrients, are also strongly affected by soil pH (Jaillard et al. 2003, Paul 2007). Consequently, major soil ecosystem services depend on pH, such as the usability for agricultural and forest production, the regulation of nutrient cycles, storage and availability, the filtering, buffering and transformation of chemicals, and the suitability of soils as a habitat and reservoir of biodiversity.

The pH is defined as the negative decadic logarithm of the H^+ ion or more exactly of the hydronium ion (H_3O^+) activity.

$$pH = \log_{10}(1/H^+) = -\log_{10}(H^+) \tag{3.1}$$

Hydronium ions, or we may simply say H^+, are formed by the dissociation of water molecules:

$$2\ H_2O \rightleftharpoons (H_3O)^+ + OH^- \tag{3.2}$$

$$H_2O \rightleftharpoons H^+ + OH^- \tag{3.3}$$

The equilibrium reaction (Eq. 3.3) can be written using the law of mass action, which relates the activity of ions formed by the dissociation of water to the activity of undissociated water:

$$(H^+) \times (OH^-) / (H_2O) = K \tag{3.4}$$

Because the activity of water in dilute solutions is almost constant at 1, the equilibrium constant K simply depends on the product $(H^+) \times (OH^-)$, which has a value of 10^{-14} for pure water at 21 °C.

$$K = (H^+) \times (OH^-) = 10^{-14} \tag{3.5}$$

This means that the activities of H^+ and OH^- are linked to each other; as H^+ increases, OH^- declines and vice versa. Due to this relationship the German standard DIN 19260 (DIN 2012) defines that pH can be only measured in aqueous solutions and is in a range of 0 to 14 (Degner 2012). Yet, pH can also exceed that numerical range. Typical pH values in soils, however, are in a range between pH 2.5 and 11; pH values in soils of humid regions are even mostly in a narrower range between pH 4.5 and 7.5 (Fig. 3.1). Hydroxides of Al and Fe become instable at pH values below 4.5 and 3.0, respectively, so that Al^{3+} and Fe^{3+} are strongly released to the soil solution when the pH is dropping below these tipping points (Parker 2005). Some soil characters and qualifiers, respectively, are assigned to soil pH determined in water and $CaCl_2$, respectively (IUSS Working Group WRB 2014), as outlined in Table 3.1.

Table 3.1. Assessment of soil properties based on soil pH and organic matter (OM) content (IUSS Working Group WRB 2014).

pH in	pH value	OM	Character/Qualifier	Properties
H_2O	> 8.7		ultrabasic (non-calcaric)	Na_2CO_3, $MgCO_3$
	8.0–8.7		basic (calcaric)	$CaCO_3$
0.01 M $CaCl_2$	< 5.1	> 15%	dystric	base saturation < 50%
	< 4.6	4–15%		
	< 4.2	< 4%		
	< 3.6	> 15%	hyperalic	base saturation < 10%, high Al-saturation
	< 3.4	4–15%		
	< 3.2	< 4%		

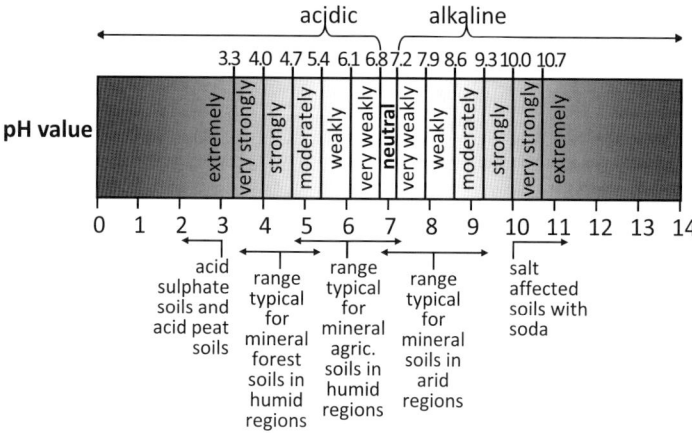

Fig. 3.1. Classification of soil pH (determined in 0.01 M $CaCl_2$) according to German soil mapping guideline KA 5 (ad-hoc AG Boden 2005) and typical pH ranges of soils.

The pH values of soil show spatial heterogeneity so that they may vary considerable within a field or plot. Additionally, pH values usually vary within a soil profile and even within soil microhabitats, such as earthworm burrows and the rhizosphere (Hawkes et al. 2007, Tiunov and Scheu 1999). Furthermore, pH also changes with time. Small, short- to mid-term variations on a scale of minutes to months are caused by variations of soil temperature, Ca^{2+} activity, and CO_2 partial pressure in soil air and soil solution. Variations are also caused by acid/base reactions related to changes in the redox-potential, biological activities of plant roots and soil organisms, and also human activities such as the application of lime and fertilizers (Blume et al. 2011, Tan 2011). Long-term effects, possibly resulting in stronger shifts in soil pH that follow a trend, may be the cumulative result of repeated short-term effects and may be caused by, e.g. mineral weathering and the immission of acid rain and atmospheric nitrogen deposition (Blake 2005, Rengel 2003). Typically, pH values of soils under temperate climate progressively decline in the long term, a process which has been substantially accelerated in many temperate forest soils due to air pollution with SO_2, NO_x, and NH_3 and resulting immissions into the soil (Alewell 2003, Blake 2005, Schimming 2011).

Consequently, (i) mapping the spatial heterogeneity of pH, e.g. within an agricultural field, and (ii) a soil depth- and time-resolved in-situ determination of pH is required to monitor the conditions for biological and chemical processes in soil. For example, pH strongly controls the mobility, bioavailability, and transport of nutrients and potentially toxic elements (e.g., Cd, Pb, Hg) in soil. The optimum requirement for phosphate fertilizer application to an arable soil may spatially vary with pH, and thus requires pH-mapping for precision farming. When, in another example, sewage sludge that is contaminated with cadmium is used as fertilizer, soil pH and cadmium mobility will be temporarily strongly affected by the sludge and will vary among soil horizons. The pH measurement of homogenized grab samples using standard laboratory methods would not be sufficient to assess the exemplarily outlined problems.

Until today, the standard measurement of soil pH is not done in the field but conducted in the laboratory under defined boundary conditions, using air-dried, sieved (< 2 mm) and homogenized soil samples. Thereby, the soil structure (small-scale heterogeneity), temperature, soil water content, and the CO_2 partial pressure, redox potential and biological activities are strongly modified, and this has substantial implications for the pH value obtained (Elberling and Matthiesen 2007). Anyhow, this is accepted due to significant problems that were previously reported for the in-situ measurement of pH in field soil using common electrometric calomel-glass reference electrodes (for description of the electrodes see section 3.1.2).

Technical problems may arise under field conditions with the use of common glass electrodes due to losses of the inner aqueous electrolyte solution to unsaturated soil and drying of the glass membrane (Schaller and Fischer 1981) as well as possible frost damage to the electrode. Even more, the suspension effect (SE) may unpredictably alter the in-situ pH reading (Al-Busaidi et al. 2005, Oman 2000a, Oman 2000b, Pallmann 1930). The SE results in different pH readings, when pH is measured in the soil (sediment) compared to the equilibrium solution (supernatant) of a settled soil suspension (Oman et al. 2007, Overbeek 1953) (see section 3.1.2 for further information on SE).

These problems have been operationally solved by the convention to measure the pH in the supernatant of a settled soil suspension using either deionized H_2O, or aqueous solutions of KCl or $CaCl_2$ with different defined soil-to-solution ratios (see section 3.1). However, all these methods only allow for short-term, individual measurements and are unsuited for longer-term monitoring and fast "on-the-go" mapping of soil pH (Schirrmann et al. 2011, Soriano-Disla et al. 2014, Thiele-Bruhn et al. 2015).

Consequently, alternative methods for pH measurement are needed and have been proposed to enable in-situ monitoring of pH in soils. Instead of using glass electrodes with Ag/AgCl system alternative methods have been proposed, i.e. more robust antimony (Sb) electrodes (Jaillard et al. 2003, Schaller and Fischer 1981), non-glass chemical sensor electrodes based on semiconductor technique termed ion sensitive field effect transistors (ISFET) (Matthiesen 2004), tin oxide (SnO_2) transparent electrodes (Tamogami et al. 2009), composite electrodes based on quinhydrone (Kahlert et al. 2004), and colorimetry based on chemical pH indicator dyes. The pH-dependent dye color is determined using optical sensors or photometry. As optical sensor devices, for example, polymer optical fibers (Blossfeld and Gansert 2007, Motellier et al. 1995),

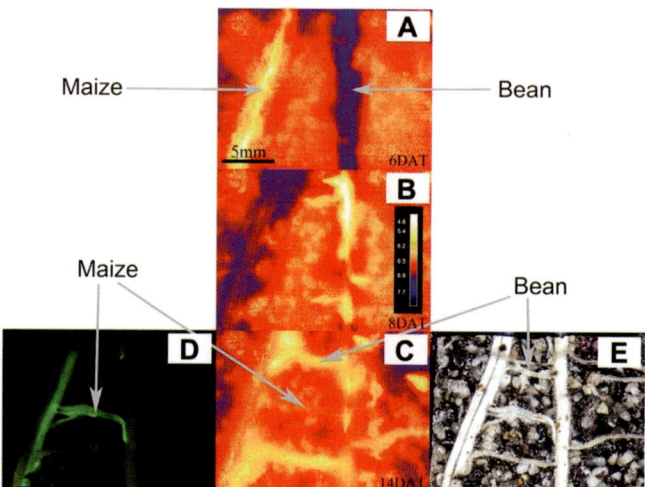

Fig. 3.2. Panels **(A–C)** show the pH maps of pH measured with an optode for the rhizosphere and adjacent soil of roots of maize and bean growing in close proximity. Recorded at a scale ranging from 4.6 to 7.7 pH units at **(A)** 6, **(B)** 8, and **(C)** 14 days after transplanting seedlings of comparable size into rhizotrons. Panels **(D, E)** show photographs of these same regions taken on day 14 after having removed the optode for locating and identifying roots. Panel **(D)** was photographed under blue light to excite the maize expressing the green fluorescent protein, allowing the identification and exact location of the maize roots. Panel **(E)** shows a conventional photograph that is complementary to **(D)**, where all the roots from maize and bean are visible. (Figure and modified text from Faget et al. (2013). (Reproduced with permission from the authors).

luminescent planar optodes (Schröder et al. 2007, Stahl et al. 2006), and optodes with fluorescent detection system (Faget et al. 2013), as well as fluorescence alterations of pH-sensitive fluorophores (Boldt et al. 2004) have been developed for real-time and/or in-situ pH measurement. An example for the results of such a measurement is shown in Fig. 3.2.

Optical methods are very good for small-scale resolution, two-dimensional pH recording and imaging (Fig. 3.2). They can be ideally combined with experimental set-ups that enable to observe soil surfaces such as rhizotrones and rhizoboxes. At the same time their use is restricted to measurements on a sample's surface (Blossfeld and Gansert 2007, Boldt et al. 2004, Stahl et al. 2006), e.g., the rhizoplane (Blossfeld et al. 2011, Schreiber et al. 2012). Optical methods are not intended to be used for autonomous field monitoring underneath the soil surface.

Sb electrodes are much less susceptible to soil drying and are mechanically much more robust than glass electrodes on the one hand. On the other hand, Sb electrodes have substantial disadvantages. Their pH working range is rather limited (Jaillard et al. 2003) and the pH measurement is influenced by complexing agents such as oxalate and citrate (Conkling and Blanchar 1988) and by several metal cations such as Cu^{2+} (Schaller and Fischer 1981). Even more relevant is that the pH measured with Sb electrodes varies with soil water content because the measurement depends on the partial pressure of O_2. Consequently, pH readings using Sb electrodes depend on the redox potential of a soil (Jaillard et al. 2003) and might, thus be biased by in-field measurements, where redox potentials are expected to change.

The ISFET electrode based on semiconductor technique can be mounted on a soil corer in order to determine pH along entire soil profiles (Matthiesen 2004). The data so obtained are reliable but the ISFET electrode is not suitable for permanent installation, a prerequisite for long-term soil monitoring (Matthiesen 2004). The ISFET electrode and other electrodes such as matrix electrodes and the mentioned metal oxide electrodes all have limited pH working ranges and shorter lifetimes than the glass electrode (Jaillard et al. 2003). For the mapping of soil pH by on-the-go systems, the mechanically more robust Sb or ISFET electrodes are often used, despite the outlined limitations. Furthermore, an Sb electrode has been proposed for use in an automated laboratory device for soil pH analysis (Decker et al. 2017). Composite quinhydrone electrodes have been reported to be suited for in-situ pH analysis under varying soil water content status (Kahlert et al. 2004). However, further experiences with that technique are lacking.

In order to simultaneously monitor soil pH and redox potential autonomously (see chapter 2 of this book), a system for the continuous determination of pH in-situ and in unsaturated soils has been developed. This system uses improved glass electrodes (Thiele-Bruhn et al. 2015) and will be presented in the following sections.

3.1.2. Fundamentals

Technical principles of electrometric pH measurement

The most often used technique is the electrometric pH measurement using a glass electrode. Standardized procedures are typically based on this technique. Electrometric sensors consist of a measurement electrode and a reference electrode. Glass electrodes have a pH sensitive silicate glass membrane, which is the most powerful and widely applicable pH sensor (Degner 2012). The interface between solution and the thin glass membrane (about 50 µm thickness) is the measuring electrode in its strict sense (Fig. 3.3). Proton transfer occurs from the solution to the

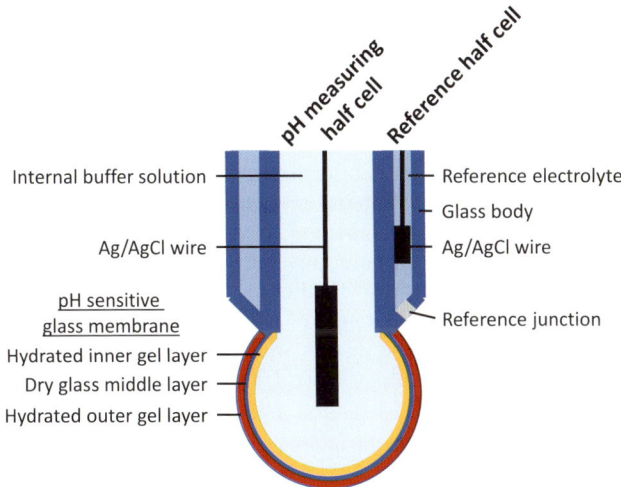

Fig. 3.3. Schematic figure of a combination glass electrode.

surface of the glass membrane and vice versa. This transfer affects the charge of the membrane surface, leading to movement of freely movable sodium and lithium cations in the silica glass, and thus altering the electric voltage of the measuring chain (Baucke 1994a, Baucke 1994b).

The glass membrane consists of about 70% silicon dioxide (SiO_2), substantial proportions of alkali metal, and alkaline earth metal oxides. The inner electrolyte solution is most often a neutral buffer solution such as 3 M KCl solution, alternatively supersaturated solutions or gels serve as reference electrolyte (Galster 1990). Gels have some advantages compared to liquids, as gels lose KCl more slowly than an electrolyte solution (0.1 µmol h^{-1} compared to 6 µmol h^{-1} for electrode with electrolyte solution). Additionally, gels can take pressure of several cm water head without ingression of water into the reference electrode (Degner 2012). Furthermore, the desiccation of gels is reduced by additions of ethylene glycol, glycerin (Galster 1990), or agar gel (see section 3.4.2.1). On the other hand, gels have the disadvantage that diffusion of contaminating ions and exchange with K$^+$ is much higher compared to electrolyte solution. This possibly leads to interferences.

Activity

When measuring pH, the hydronium ion activity is determined. Consequently, solutions with similar solute concentrations may have a different pH value because the hydronium ion activity not only depends on the hydronium ion concentration but also on the mobility of hydronium ions. This mobility hinges on the total ion concentration of a solution, the type of dissolved ions and the temperature (Bühler and Bucher 1982, DIN 2000).

The activity can be approximately calculated using the Debye-Hückel equation for determination of the activity coefficient f_i in very dilute solutions (Suarez 1998):

$$\log f_i = -A \times z_i^2 \times I^{0.5} \qquad (3.6),$$

where z_i is the charge of the ion in question, I the ionic strength of the solution, and A is a constant, which itself depends on the dielectric constant, temperature, and density of the solution. For solutions with concentrations of $I > 0.01$, extended forms of the Debye-Hückel equation must be used that have additional parameters. At 25 °C A has a value of around 0.51. The activity a is calculated as

$$a_i = f_i \times c_i \qquad (3.7)$$

As numerous parameters such as temperature influence ion activity, it is obvious that pH is not an absolute value but depends on environmental and measuring conditions. According to German standard DIN 19260 (DIN 2012) pH measurement is only applicable for aqueous liquids within a range of pH 0 to 14 and in dilute aqueous solutions with a total concentration of $c \leq 1$ mol l^{-1}. When measuring pH in practice, each of these conditions may possibly not be met. Yet, for soil it is not expected that pH values exceed the range of 0–14, the concentration in soil solution is typically ≤ 1 mol l^{-1}, and nonaqueous phase liquids are largely restricted to contaminated industrial sites where accumulations, e.g., of oil, tar or BTEX aromatics, may occur.

Measuring pH under controlled conditions

When pH is determined with the most commonly used glass electrode, an electric potential is measured. The voltage (E) at the pH electrode is the sum of six individual potentials (see Fig. 3.4) (Galster 1990):

- the potential at the conduction system of the glass electrode (E_1),
- the potential at the inner surface of the glass membrane (E_2),
- the asymmetry potential of the glass membrane (E_3),
- the potential at the outer surface of the glass membrane (E_4),
- the diffusion potential at the diaphragm (E_5),
- and the potential of the reference element of the reference electrode (E_6),

so that

$$E = E_1 + E_2 + E_3 + E_4 + E_5 + E_6 \text{ (mV)} \qquad (3.8)$$

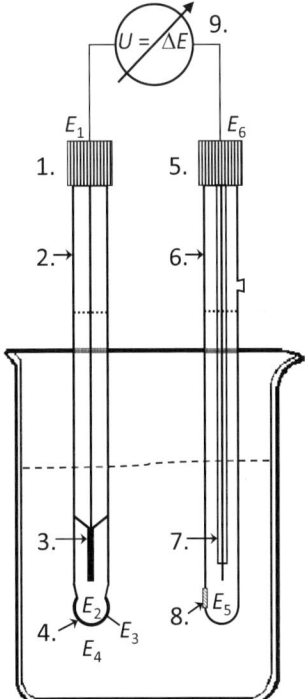

Fig. 3.4. Schematic design of a two-probe glass electrode system with a glass measuring electrode (1), shaft (2), inner reference system with Ag wire inserted in electrolyte (3), hydronium ion sensitive glass membrane (4), reference electrode (5), shaft with opening for filling in KCl electrolyte (6), outer reference system with metal wire (Pt or Ag) inserted in KCl electrolyte (7), the porous diaphragm (8), and electrical connection and measuring device (9). See text for explanation of individual potentials $E_1 - E_6$. (Redrawn from Calmano (without date)).

Steepness

The change in voltage per pH unit is reflected by the steepness. At room temperature of 20 °C a theoretical absolute value of 58.2 mV is obtained from the Nernst equation:

$$E = E_0 + (R \times T)/(n \times F) \times \ln(H^+) \qquad (3.9)$$

with the electrode potential E, the standard potential E_0, R the gas constant (8.314 J mol^{-1} K^{-1}), T the absolute temperature in Kelvin, n the valence of the ion considered (valence of H$^+$ is 1), F the Faraday constant (96.487 °C mol^{-1}), and the activity of the hydronium ion (H$^+$). From Eq. 3.9 it becomes clear that the change in voltage per pH unit, which is the steepness, is temperature dependent. If the effective steepness (S_{eff}) of an electrode is known, the steepness at temperature ϑ (°C) can be calculated from Eq. 3.10:

$$S_\vartheta = S_{eff} \times (T+\vartheta)/(T+1) \qquad (3.10)$$

with S_ϑ the steepness at temperature ϑ (°C), and the temperature T (K) at which S_{eff} was determined.

This temperature dependence results in shifts of the steepness of the pH electrode from 54.20 mV per pH unit at 0 °C to 74.04 mV per pH unit at 100 °C. In most customary pH meters, the change in the steepness is compensated in a range around room temperature (Degner 2012). However, a single general temperature compensation factor is not sufficient to compensate the temperature dependent change in the steepness, because the relation of temperature and steepness also depends on the specific electrode and the matrix or solution in which the pH is measured. Temperature affects the calibration data of the offset voltage and slope, so that changes in temperature to some extent bias the pH measurement. In case a zero point is as typically set at pH = 7, the slope value is 0 and remains unaffected by shifts in temperature. Yet, a temperature-dependent deviation occurs and linearly increases under more acidic and basic conditions, respectively. Imagine soils that cover the typical pH range from 3 to 8 at room temperature and are exposed to a change in temperature of ± 20 °C. The maximum change in pH (increase and decrease, respectively) would occur at the combination of extreme temperatures (0 °C, 40 °C) and extreme pH value (pH 3). Under these extremes a maximum deviation of ±0.27 pH units would result. These technical aspects are good reasons, why pH is measured in the laboratory under controlled conditions at temperatures usually adjusted at 20 to 25 °C. This enables to receive pH data that are fully comparable between samples (Degner 2012).

Even more, temperature affects not only the pH measurement but also pH itself such as the H^+ ion activity as a thermodynamic entity. Moreover, properties of the measuring solution, i.e. solubility of ions, viscosity of the solution, dissociation of dissolved substances, and biological activity are influenced by temperature. This is relevant in a field situation but excluded in standardized laboratory procedures. For example, the dissociation, and thus the pH of water, changes from pH 7.00 at 25 °C to pH 7.47 at 0 °C (Bates 1973).

Soil temperatures in the temperate climate zone are largely in the range of 20 ± 20 °C. Temperatures outside this range may be reached under specific weather conditions at and near the soil surface, in case soils are not covered by vegetation or snow (Bachmann 1997, Lahl et al. 2012). Also keeping in mind that agricultural soils often have pH values in a narrower range of, e.g. 4.5 to 7.5, the possible technical error due to inadequate compensation of the temperature effect on the steepness is rather negligible. In many cases the temperature effect on pH will even be within or close to the measurement accuracy of pH electrodes of ±0.05 pH units (Mettler-Toledo 2007). These are reasons why the here recommended field pH measurement is carried out without temperature compensation.

Sources of errors and abrasion
Suspension effect

The suspension effect (SE) may unpredictably alter the pH reading (Al-Busaidi et al. 2005, Oman 2000a, Oman 2000b, Pallmann 1930). The SE results in different pH readings, when pH is measured in the soil (sediment) compared to the equilibrium solution (supernatant) of a settled soil suspension (Oman et al. 2007, Overbeek 1953). The SE is based on two partial effects. The first, SE1 occurs because non-dissolved (particulate and dispersed) components are able to polarize water molecules in their vicinity, and thereby, increase the activity of hydronium and hydroxide ions in solution (Degner 2012). The pH change increases almost linearly with the concentration of dispersed particles and may reach ΔpH values of up to 2 (Schwabe 1976). The second effect, SE2 results from KCl solution flowing or diffusing from the salt bridge of the reference electrode, when it is directly inserted in the soil. When KCl gets in contact with soil particles, it causes an abnormal liquid junction potential at the particles/KCl solution interface (Oman 2012). It is assumed that this effect should be particularly strong in dry soil lacking a substantial dilution of KCl with the measuring solution and/or soil solution (Davis 1943).

Interference from alkali ions

The potential at the membrane of the glass electrode depends not only on the hydronium ion activity (pH) but also on the alkali ion activity (pM). This is especially the case for Li^+ and Na^+, while K^+, Cs^+, and Rb^+ have practically no influence on pM (Baucke 1994a, Baucke 1994b, Covington et al. 1985). Because lithium is not present in soils in relevant concentrations, the alkali effect on the membrane potential (alkali error) is practically a sodium error. Depending on the sodium concentration, it leads to a charge displacement and erroneously lowered pH readings. This effect is not relevant at pH below 9 though. Consequently, the alkali error is restricted to salt and sodium rich alkaline soils such as Solonchak and Solonetz. Technically, the problem can be circumvented by using lithium rich membranes with low cross-sensitivity to alkali ions (Degner 2012).

The exchange of ions with the soil solution affects not only the surface of the electrode membrane but also the underlying glass layer and may lead to degradation of the glass. However, this process is least effective in the pH range 3 to 5 where it would take more than 100 years to dissolve 0.1 mm of membrane. Typically, glass membranes are very durable, and at 25 °C, they have a lifetime of several years (Covington et al. 1985).

Loss of electrolyte from solution (or gel) by outflow from the inside of the reference electrode into the outer solution may affect the lifetime of the electrode even more. Yet, this outflow cannot be completely prevented, but a small outflow is even required as it keeps soil solution (measuring solution) from penetrating into the electrode. For this purpose the reference electrode is equipped with a transfer opening, called reference junction.

A too high loss of electrolyte may produce an undesirable diffusion voltage, when anions and cations have a different rate of diffusion. For that reason, KCl is typically used as salt in electrodes, because the diffusion rates of K^+ and Cl^- are rather similar.

Dirty electrode

When measuring pH in soil, it can hardly be avoided that the electrode becomes dirty from adhering soil. However, this may have several effects. A dirty glass membrane reacts more slowly to pH changes, so that short duration pH readings may be erroneous. A contaminated diaphragm may also cause memory effects and irregular temperature behavior, which will also affect the measured pH. This is relevant if a glass electrode is used for multiple, sequential short-term pH measurements (Degner 2012). Yet, these problems are irrelevant for long-term installed electrodes (pH monitoring), where no abrupt, but only gradual changes of pH and of the composition of the soil solution are expected and continuously recorded.

Hardware damage

Special care has to be taken when installing and handling the glass electrode. Although, glass electrodes designed for soil pH measurement are quite robust, the glass membrane may be easily damaged. This must be considered especially when installing a glass measuring electrode in soil (see section 3.4.2.1).

The pH measuring instrument requires a power supply. A broken electric cable to the electrode will cause an extremely high electrical resistance. This may result in very unstable pH readings and subsequent overflow of the pH meter (Degner 2012).

3.2. Method selection

Methods for soil pH measurement described in guidelines and handbooks are typically meant to be carried out under controlled laboratory conditions. A selection of some of these methods is listed in section 3.3. They should be used where pH is to be determined as a general soil property, with optimal precision, and to obtain pH data that are fully comparable between different soil samples, sites, and studies.

Short-term field measurements of grab samples using colorimetric methods or hand-held pH meters are meant to obtain quick, rough pH information. Data should be rounded to one digit or less.

Continuous pH monitoring in the field with long-term installed electrodes yields temporally resolved data on pH at actual environmental and soil conditions. For being able to properly assess such data, pH monitoring should be combined with the monitoring of redox potential, soil moisture, and soil temperature.

Spatially resolved pH information can be obtained by using on-the-go soil pH sensors as an alternative to the laborious field soil sampling and subsequent pH analysis in the laboratory. A further option is the use of spectroscopy with visible and near-infrared light (VNIRS) or mid-infrared light (MIRS) (Brown et al. 2006, Chang et al. 2001, McCarty and Reeves III 2006, Pirie et al. 2005, Viscarra Rossel et al. 2006). Airborne or handheld instruments are available for that purpose, the latter sensors can even be mounted on an on-the-go system. However, it must be regarded that spectroscopy delivers indirect measurements from which soil properties such as organic carbon content and also pH can only be derived by applying statistical models. At present, the evaluation of spectroscopic data requires careful calibration using large calibration data sets and specific estimation models for data obtained at the investigated site or at least for soil sets with similar properties as the site to be examined (Vohland et al. 2016). On top of that, it must be noted that pH, other than e.g. soil organic matter, is a soil property that does not directly affect the spectroscopic properties of soils. Hence, determining pH with spectroscopic methods yields only an indirect estimation.

3.3. Laboratory methods for measuring pH in soils

Measuring soil pH in the laboratory under controlled conditions is still the standard procedure, although, it has significant drawbacks compared to a direct measurement in the field. Due to this, some of the major methods that are used in many countries are described in the following and options to compare the operationally defined pH-readings are given.

Determination of pH using standard laboratory methods requires the destructive sampling of soil by taking a sample in the field. Subsequently, the soil samples are typically air-dried, sieved (≤ 2 mm) and homogenized; analyses are carried out at room temperature. As mentioned before, such destructive sampling will inevitably change soil properties relevant for pH such as redox potential, CO_2 partial pressure, and biological activities. Depending on the aims of the study, sampling may be done with high spatial or vertical resolution, e.g., soil from different soil horizons or microhabitats such as the rhizosphere. Yet, spatial resolution is practically limited, because several grams of soil are required for the determination of pH in the laboratory.

The standard methods listed below are recommended for mineral soils. Especially, the suggested soil-to-solution ratios may not be suitable for organic soils and organic layers with very high water holding capacity.

The most commonly used solutions for pH measurement in the laboratory are distilled H_2O, 0.01 M or 0.1 M calcium chloride ($CaCl_2$), and 1 M potassium chloride (KCl) solutions. Typical soil-to-solution ratios are 1:1, 1:2, 1:2.5, 1:5, and 1:10 (w/w) (Al-Busaidi and Cookson 2002,

Blume et al. 2011, Degner 2012, DIN 2005, Jahn et al. 2006, Tan 2011). In most methods, the solutions are used to prepare suspensions from air-dried soil samples, in which pH is measured. Some methods also recommend the separation of the soil solid phase by centrifuging, decanting, and/or filtration (Tan 2011).

In the field, soil samples are often suspended in H_2O because it is available almost everywhere (being possibly not distilled H_2O). The method is suitable for detecting seasonal and even diurnal variations of soil pH; the pH_{H2O} is also termed as "actual acidity". Salt solutions are used for suspension in order to receive more stable pH readings, with reduced temporal variation in soil pH. The $CaCl_2$ solution provides Ca^{2+} ions that displace other ions, especially H^+ ions from exchange sites. The resulting pH value, termed as "exchange acidity", is less affected by a) seasonal changes and b) by the concentration of other soluble salts in the suspension. The $CaCl_2$ method is best suited to assess the soil reaction of agricultural soils. The KCl solution is especially used for pH measurement of acidic soils. Hence, it is the preferred method for analyzing pH in soils of temperate forests. The higher concentrated KCl solution (typically 1 M) displaces H^+ and Al^{3+} completely from the exchange sites, which is not always the case when more dilute 0.01 M $CaCl_2$ solution is used. The displaced Al^{3+} then consumes OH^- ions, increases H^+, and thus lowers pH. Generally, exchangeable aluminum is present at pH $\leq$ 5.2 and the Al^{3+} concentration largely increases at pH < 4.5. Above pH 5.2, aluminum is non-exchangeable because of hydrolysis, polymerization, and precipitation. The absolute value of pH_{KCl} is strongly correlated with aluminum saturation. However, it must be noted that 1 M KCl substantially alters the original soil conditions. The pH_{KCl}, termed as "potential acidity", is unsuitable for interpreting a soil's fertility or crop production potential.

The pH meter should be calibrated using two pH buffer standards that cover the pH range expected for the soils analyzed, i.e., pH 2.0 and 4.0, pH 4.0 and 7.0, pH 7.0 and 10.0 (Hendershot et al. 2006).

Selected guidelines for pH analysis of soil

DIN 19684-1 (1977) Methods of soil analysis for water management for agricultural purposes; chemical laboratory tests; determination of the pH value of the soil and the lime requirements

To analyze mineral soil samples, 10 g soil are mixed with 25 ml 0.01 mol l^{-1} $CaCl_2$, stirred and left for $\geq$ 1 h. Measurement with the glass electrode is done in the settling soil suspension. Stability criterion: $\Delta V \leq 1$ mV/30 sec that is the measured voltage (V) must not vary by more than 1 mV within 30 seconds.

To measure the pH in peat soil, a 1:3 (volume fraction) suspension is prepared, using field moist soil.

The pH recordings are reported rounded to two decimal digits (e.g., pH = 6.28).

ISO/DIN 10390 (DIN 2005) Soil quality – Determination of pH

ISO 10390:2005 specifies an instrumental method for the routine determination of pH using a glass electrode in a 1:5 (typically 10 g soil with 50 ml solution) suspension of soil in water (pH in H_2O), in 1 mol l^{-1} KCl solution (pH_{KCl}) or in 0.01 mol l^{-1} $CaCl_2$ solution (pH_{CaCl2}). The soil suspension is stirred and allowed to equilibrate for 2 to 24 h. The suspension is stirred up again and its pH is measured in the settling soil suspension. Stability criterion: $\Delta pH/t \leq 0.02$ pH units per 5 seconds.

The pH recordings are rounded to two decimal digits.

It is important that the measurements should comply with the stability criterion. The time required to reach a stable and reliable pH reading increases with pH (logarithmically declining H^+ activity) and may take several hours for alkaline soil samples.

In most soils and studies measured pH values decrease in the sequence $pH_{H2O} \geq pH_{CaCl2} \geq pH_{KCl}$ (Gascho et al. 1996), and measured pH is higher at wider soil-to-solution ratios (e.g., 1:5) compared to closer ratios (e.g., 1:1). Yet, the pH data obtained in H_2O, $CaCl_2$, and KCl suspension are highly correlated and lie closely together as is shown on example of the following regression equations.

- $pH_{H2O(1:5)} = 1.06 \times pH_{H2O(1:1)} + 0.52$
  $R^2 = 0.91$, $n = 563$ mineral soil horizons and Histosols, $pH_{H2O(1:5)}$ range: 3.4 to 10.5 (Libohova et al. 2014).

- $pH_{H2O(1:5)} = 1.01 \times pH_{CaCl2(1:2)} + 0.41$
  $R^2 = 0.92$, $n = 563$ mineral soil horizons and Histosols, pH_{CaCl2} range: 2.9 to 10.5 (Libohova et al. 2014).

- $pH_{H2O} = 1.03 \times pH_{CaCl2} + 0.40$
  $R^2 = 0.97$, $n = 94$, soil:solution 1:5, mineral soil horizons, pH_{H2O} range: 5.97 to 8.97 (Gavriloaiei 2012).

- $pH_{KCl} = 1.007 \times pH_{H2O} - 0.642$
  $R^2 = 0.98$, $n = 94$, soil:solution 1:5, mineral soil horizons, pH_{H2O} range: 5.97 to 8.97 (Gavriloaiei 2012).

- $pH_{KCl} = 1.080 \times pH_{CaCl2} - 0.400$
  $R^2 = 0.97$, $n = 45$, soil:solution 1:2.5, forest soils and litter layers, pH_{CaCl2} range: 2.59 to 6.16 (Koch and Thiele-Bruhn, unpublished)

The linear correlation of pH measured in 0.01 M $CaCl_2$ and 1 M KCl is exemplarily illustrated in Fig. 3.5. It has been reported, though, that the relationship between pH_{H2O} and pH_{CaCl2} is not linear over the entire pH range that is typically found in soils (Ahern et al. 1995, Aitken and

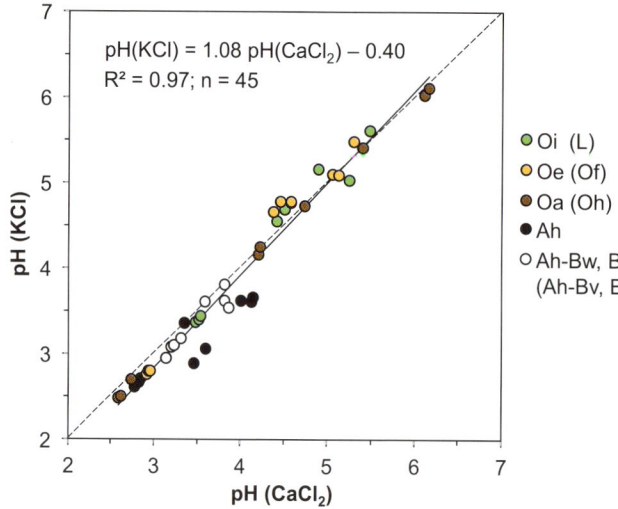

Fig. 3.5. Relationship of soil pH determined in 0.01 M $CaCl_2$ and 1 M KCl suspension, respectively; samples from different mineral horizons and litter layers of forest soils (horizon nomenclature according to WRB and German KA5 classification in parentheses). The dashed line indicates the 1:1 ratio, the regression is indicated by the solid line. (Koch and Thiele-Bruhn, unpublished)

Moody 1991, Henderson and Bui 2002). Consequently, other, nonlinear (polynomial) calibration equations have been proposed. For practical purposes, however, it seems reasonable to use the more parsimonic linear models (Little 1992, Miller and Kissel 2010, Minasny et al. 2011).

Several methods for measuring soil pH are described in the Soil Survey Laboratory Methods Manual of the United States Department of Agriculture (Soil Survey Staff 2004). Those methods relate to criteria in the Soil Taxonomy (Soil Survey Staff 2010). Many of them are closely related to the DIN and ISO methods described above and are, thus, not presented in all details. Suspensions with H_2O (soil-to-solution ratio 1:1), 0.01 M $CaCl_2$ (soil-to-solution ratio 1:2), and 1 M KCl (soil-to-solution ratio 1:1) are recommended as described above.

Additionally, there are a number of methods to identify specific pH-related soil properties. Of this, three are presented in the following.

Delta pH

The numerical difference in the values of pH measured in KCl and H_2O is referred to as the delta pH. This difference enables to assess the net charge at a soil's exchange sites. When delta pH is negative, the soil colloids have a net negative charge and vice versa at a positive delta pH. For example, highly weathered Oxisols (Ferralsols according to WRB) are characterized by a net positive charge and a high anion-exchange capacity. If the content of organic matter is additionally low, Cl^- ions from KCl displace OH^- ions from the anion exchange sites, resulting in a higher pH compared to that determined in H_2O. Delta pH is not considered when the pH is higher than about 6.5 (Soil Survey Staff 2004).

Oxidation pH

Oxidation pH is used for determining the presence of sulfidic material and of sulfuric horizons, respectively. Sulfide minerals, such as pyrite, and/or elemental sulfur in reduced sulfidic sediments or mine tailings oxidize upon exposure to air, yielding sulfuric acid. In the laboratory procedure to determine the oxidation pH, the natural, microbial formation of acid sulfate is accelerated by incubating a water-saturated soil sample in a closed container at room temperature. The O_2 needed for the oxidation process is supplied by periodic stirring of the sample. The sample is given sufficient time of up to 8 weeks to oxidize completely, before a pH_{H2O} (1:1) is measured. Stability criterion: $\Delta pH \leq 0.03$ pH units before versus after stirring and settlement. Hydrogen peroxide may be added to the soil in order to accelerate oxidation. Any sulfidic material present will result in violent effervescence and an extremely acid suspension (Soil Survey Staff 2004).

pH measurement in 1 M sodium fluoride (NaF)

Measuring pH of a suspension containing 1 g soil in 50 mL 1 M NaF is used to test for andic properties in soils, i.e. allophane and/or organo-aluminum complexes are present (Fieldes and Perrott 1966). This is typically the case when pH $\geq$ 9.5 is reached within a few minutes in NaF suspension (according to WRB; in comparison, a pH_{NaF} of 8.4 is required according to USDA). The action of 1 M NaF on these young pyroclastic minerals releases hydroxide ions (OH^-) into the soil solution and increases its pH. The test works in most layers with andic properties, except of those very rich in organic matter. As a bias, however, the same reaction also occurs in spodic horizons and in certain acid clays rich in Al-interlayered clay minerals. The test fails in calcareous soils (Jahn et al. 2006, Soil Survey Staff 2004).

3.4. Field methods of determining the soil pH

3.4.1. Analysis of grab samples in the field

No specific methods exist for soil pH measurement in the field and especially not for in-situ analysis of undisturbed soils. Typically, a simplified analysis is done following the methods for laboratory pH measurement of grab samples, e.g., according to DIN 19684 or ISO 10390 (see section 3.1 of this chapter), using soil-to-solution ratios of about 1 : 1 or 1 : 2.5 and equilibration times of 30 min (Blume et al. 2011, Soil Survey Staff 2004). Portable and pocket pH meters are typically used for that purpose.

Alternatively, colorimetric methods using pH sensitive dyes may be applied. Standard dyes and the Hellige-Pehameter give a rough estimate of pH, while "non-bleeding" paper pH indicator strips give an initial assessment (accuracy of pH ± 0.5) (Blume et al. 2011).

Electric pH-meters must be regularly maintained and, when used in the field, operation in excessively hot or cold temperatures should be avoided to prevent erroneous readings (see technical errors described in section 3.1.2). Extreme temperatures and prolonged exposure to sunlight may also affect the reliability and durability of dyes. Several of the kits in use include a neutral salt, which may lead to variations in pH, when different kits have been applied. Paper pH indicator strips are less sensitive to temperature and sunlight (Soil Survey Staff 2004).

3.4.2. Continuous, in-situ pH monitoring in soil

3.4.2.1 Components of an in-situ soil pH monitoring system and installation

For continuous, in-situ pH monitoring of undisturbed field soils, a two-probe pH electrode has been re-designed especially for the permanent installation in soils (registered utility patent; Art. no. 465 (single pH electrode), ecoTech GmbH, Bonn, Germany) (Thiele-Bruhn et al. 2015). In brief, the sensor system comprises a single reference electrode, acting as the negative pole of the electrical circuit, and one or several glass measuring electrodes, acting as the positive pole. The reference probe is an Ag/AgCl-electrode (Fig. 3.6a) that is additionally embedded into a 'salt bridge' (Fig. 3.6b). The salt bridge is a tube filled with 3 M KCl (with an additive of AgCl), stabilized by agar gel in order to slow down the loss of KCl solution through the ceramic diaphragm at the tip of the reference probe. The glass measuring electrode (Fig. 3.6a) is small (diameter 6 mm) and manufactured from three different glass types to achieve a long lifetime and to prevent damaging during installation, from freezing, etc. It is equipped with a reinforced glass measuring membrane with a cylinder shape to avoid hairline cracks at the junction with the glass shaft. Additionally, it has got a transition glass section with an expansion coefficient lying between the coefficients of the membrane glass and the shaft glass to preclude damage by strong temperature changes. The electrode is filled with a 3 M KCl gel electrolyte, allowing installation in any direction. Its small diameter makes it also suitable for column and pot experiments in the laboratory; for long-term field use it can be additionally equipped with a protective shaft (Thiele-Bruhn et al. 2015).

For continuous pH monitoring, the following equipment is required: (i) glass measuring electrode, (ii) reference electrode, (iii) salt bridge, and (iv) pH meter. Instead of using a manually operated pH meter, a data logger may be used to record readings automatically.

Further information about the reference electrode is reported in section 3.1.2 of this book, on the salt bridge in section 3.1.3, on the continuous data logging in section 3.1.4.2, on the ideal number of replicates in section 3.2, and on the maintenance in section 3.6.

In order to install the glass measuring electrode, e.g. to do in-situ pH measurement in undisturbed soil, a small auger or a sharpened metal rod is used to drill a hole in the soil slightly

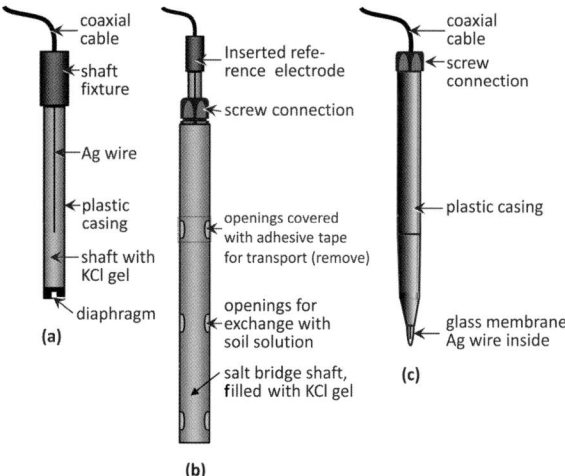

Fig. 3.6. Reference electrode (a) and salt bridge (b) with reference electrode inserted as well as measuring electrode (c) (from 'Manual for redox electrodes according to Mansfeldt'; ecoTech, Bonn, unpublished; redrawn and supplemented)

smaller than the diameter of the electrode shaft. Close contact between electrode and soil must be achieved in order to avoid creating an artificial coarse pore volume that may locally distort soil moisture conditions and soil air composition. The hole should be about 0.5 cm less deep than the depth at which the electrode is to be installed. The electrode is inserted into this hole and carefully pushed the final 0.5 cm into the undisturbed soil. In soils of high bulk density, i.e. very dry, coherent soils and/or soils with many stones and coarse roots, the depth of the installation hole should be about 1 cm deeper than the intended installation depth of the probe. The resulting void is filled with a soil paste prepared from soil of that depth, so that the glass tip of the electrode is inserted into and finally embedded by the paste. For further information on the installation of the reference electrode, salt bridge etc. in field soil see section 3.3 of this book.

Several glass measuring electrodes can be connected to a single reference electrode in a measuring unit; experiences from existing field installations have shown that distances of several decimeters between electrodes are unproblematic (Fig. 3.7) and the electrodes can even be installed at sites with different pH (Thiele-Bruhn et al. 2015). Several measuring units may be connected to one data logger. Cable connections of up to some tens of meter in length and their warming and cooling by the ambient temperature do not interfere with the correctness of the pH-readings.

3.4.2.2. Results of long-term in-situ pH monitoring in the field

Earlier studies have shown that short-term pH measurement in unsaturated soil may be highly erroneous and subject to substantial drift of the pH reading over several hours after probe installation (Davies 1943 and literature cited therein). Such error effects on the pH reading very much depend on the conditions to which the electrode had been exposed before. Consequently, long-term installation of pH electrodes is preferred, as it allows for full equilibration of the electrode to the soil conditions.

A field setup for long-term monitoring of pH is exemplary shown in Fig. 3.7. On a field site, two monitoring systems for combined long-term measurement of pH, redox potential, soil temperature, soil water content, and electric conductivity (EC) of the soil solution (salt content) had

Fig. 3.7. Systems for long-term monitoring of pH and other soil parameters (see text). Shown are the wirings and the switch cabinet that houses the data logger (Fig. 3.7a), and the pH measuring electrodes (Fig. 3.7b) that have different lengths to measure pH at three soil depths of 5, 15, and 25 cm.

been installed. Each system consisted of the respective electrodes, whereby analysis of pH and redox was done using the same reference electrode, and soil temperature, moisture, and EC were measured using combined 'Hydra probe' electrodes. The parameters were monitored on arable fields under strawberry cultivation with straw mulch (Fig. 3.7a) and with plastic foil, termed as plastic mulch (Fig. 3.7b) (Muñoz et al. 2017).

The pH values, redox potentials and soil temperatures that had been recorded over a period of four months in fall are shown in Fig. 3.8. They show clearly that pH is not a constant soil property but varied between the two mulching systems and even more within the observed time course. Variations of pH did not parallel those of the redox potential, although, both parameters are physicochemically linked, as the relation between the hydrogen partial pressure $p(H_2)$ and the derived rH value, respectively, and the redox potential (E_H) suggests:

$$rH = -\log p(H_2) = 2 \times (E^H/0.059 + pH) \tag{3.11}$$

The correlation between pH and Eh values was notably different in the two mulching systems with correlation coefficients r of -0.79 (plastic mulch) and -0.08 (straw mulch), respectively. Obviously, the soil pH was also driven by other factors such as soil temperature (correlation coefficient $r \leq 0.56$) and biologic activity; pH cannot easily be calculated from other parameters or single laboratory analyses. During the monitoring period (Fig. 3.8), the soil water content (volumetric water content) varied between 5.1% (almost air-dry soil) and 37.4% (almost field capacity). Even extreme soil water content did not lead to erroneous pH readings. This confirms prior laboratory testing, where pH measurement was disturbed only after a soil water content of 2% persisted over a four day period and extremely fluctuating and obviously erratic pH readings occurred, yet the correct measurement was reconstituted when the soil was rewetted (Thiele-Bruhn et al. 2015).

Validation of the acquired data was done at various times by parallel measurements of pH in the field and using the standard laboratory method (DIN 2005); pH values agreed very well (Muñoz and Meyer, personal communication). Hence, pH monitoring in the field can be conducted over months and even years. For example, a forest soil monitoring site of the Geological Survey of North Rhine-Westphalia, Germany, was operated for a period of three years. A subsequent calibration check in the laboratory confirmed that the calibration was still correct (Wessel-Bothe, personal communication). This proves that ageing of the electrodes did not distort the

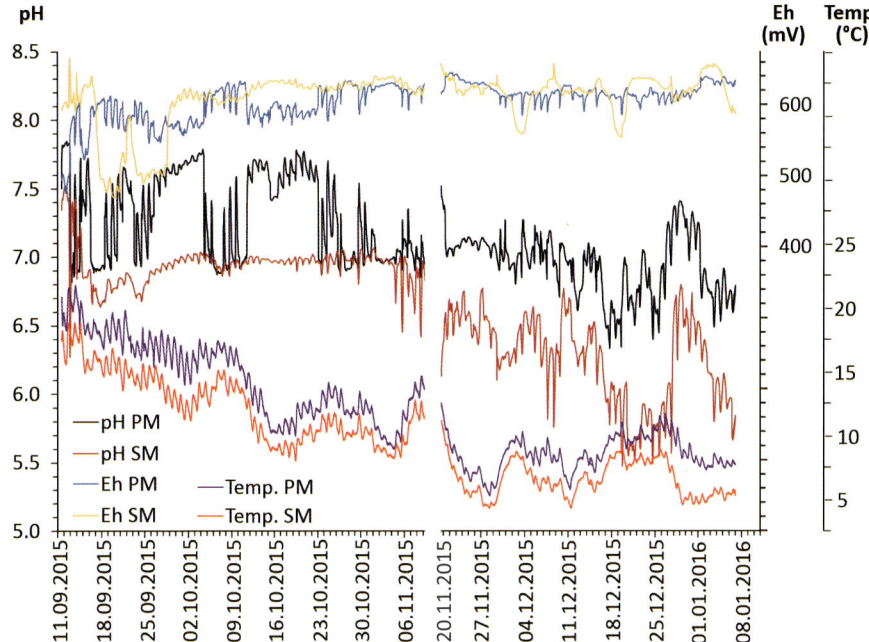

Fig. 3.8. Monitoring data of pH, redox potential (Eh), and temperature in field trials testing the effects of plastic mulch (PM) and straw mulch (SM) on soil, illustrated on the example of the period September 11, 2015 to January 8, 2016. Data are averages of independent measurements at three soil depths, i.e. 5, 15, and 25 cm. (No data were recorded from November 9 to 20, 2015).

pH measurements. Nevertheless, the calibration of such a system should be controlled from time to time (e.g. every 6 months), and the KCl liquid and gel of the reference electrode should be checked in similar intervals and replaced if necessary (see section 3.1.2 of this book).

3.4.3. Measuring on-the-go the spatial heterogeneity of pH in field soil

3.4.3.1. Automated on-the-go soil sensor coupled with soil pH mapping

Soil pH may vary considerably within an agricultural field, irrespective of the field size and even at short distance (Borchert et al. 2011, McBratney and Pringle 1997). For example, soil pH values of 17 sites had been found to vary at the field scale on average by 1 to 2 pH units (without outliers) (Olfs et al. 2012), and differences by about 2 pH units were observed within a 12 meters distance (Bianchini and Mallarino 2002). Hence, mapping spatial heterogeneity of soil pH at the scale of m² to ha is a relevant task.

To that end, traditional soil sampling (see section 3.4.3.2 of this chapter) and pH analysis in the laboratory can be done. However, the large number of samples required for this may render this approach unrealistic. Instead, on-the-go soil sensors provide an innovative technology for measuring and mapping soil properties at the field level, offering the potential for a substantial increase in measurement density at a relative low cost (Viscarra Rossel and McBratney 1997). Different tractor mounted on-the-go systems for soil pH analysis have been developed that sample the soil, prepare the samples and carry out pH measurements while traversing the field. In

Fig. 3.9 the Soil pH Manager™ on-the-go system from Veris Technologies (Salina, KS, USA) is shown. Typically, on-the-go systems comprise a soil sampling unit with a depth-adjustable (range within 0 to 20 cm soil depth), hydraulic soil sampler (No. 1 in Fig. 3.9), an optional sieving mechanism, a soil analytical component with electrodes (No. 5 in Fig. 3.9), eventually electrolytes (e.g. 0.01 M $CaCl_2$ solution), an electrode cleaning device (No. 6 in Fig. 3.9), and a digital data collection, calculation and geo-referencing unit (GPS based) possibly with online data transfer (No. 11 in Fig. 3.9), mounted on a tractor-pulled platform (Schirrmann et al. 2011, Viscarra Rossel et al. 2005). Soil tillage tools loosen the soil ahead of the sampler and break up soil crusts. A more detailed description of the Veris system can be found in Schirrmann et al. (2011). For pH analysis, glass electrodes with flat or dome surface (Adamchuk et al. 1999, Kheiralla et al. 2016, Sethuramasamyraja et al. 2008), Sb electrodes (Olfs et al. 2012, Schirrmann et al. 2011), and ISFET pH sensors (Viscarra Rossel and McBratney 1997, Viscarra Rossel and Walter 2004) are in use. In other approaches, non-invasive VIS/NIR spectrophotometer technique in combination with a tractor pulled on-the-go system was tested for pH estimation (Wang et al. 2015).

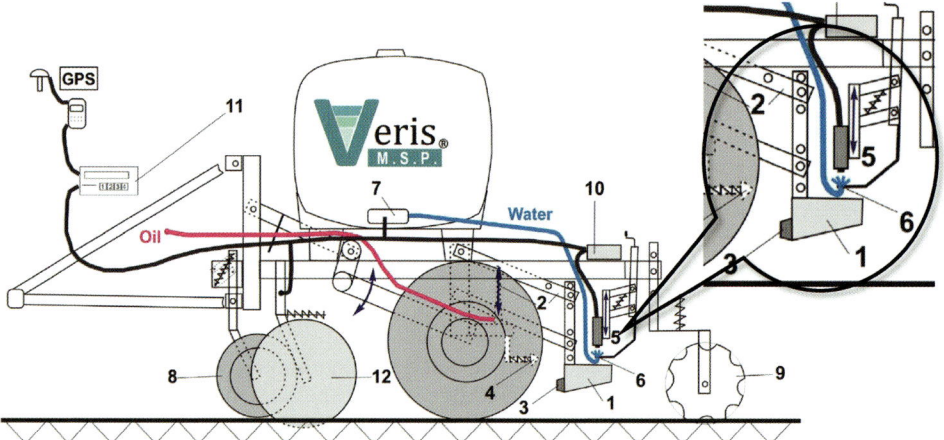

Fig. 3.9. Schematic diagram of the Veris Multi Sensor platform Soil pH Manager™ with on-the-go soil pH sensor (Source: Schirrmann et al. (2011), supplemented. Reprint with permission from the authors). The numbers shown in the figures are explained in the text.

The on-the-go systems move at a speed of 3 to 6 km/h over the field, analysing samples within a few seconds (Adamchuk et al. 1999, Schirrmann et al. 2013). Tests with Sb electrodes show more stable pH readings, when electrodes are retained in soil for more than 5 seconds; an optimum contact duration of 10 seconds was reported (Schirrmann et al. 2011). The agreement of pH data determined by the on-the-go in-field analysis and pH data obtained using standard methods under laboratory conditions is good; linear regression coefficients of $pH_{on-the-go}$ versus $pH_{standard}$ range from R^2 of 0.63 to 0.89 with no clear preference for a specific technique (Adamchuk et al. 1999, Lund et al. 2005, Olfs et al. 2012, Schirrmann et al. 2013, Schirrmann et al. 2011, Sethuramasamyraja et al. 2008, Viscarra Rossel and Walter 2004). The associated statistical error is ≤ 0.4 pH unit (Adamchuk et al. 2007, Viscarra Rossel and Walter 2004). However, to convert raw data from on-the-go analysis in the field into pH data, recalculation of the data is required using a linear regression model. In order to receive reliable data, this should not be a general model but a specific model, derived and validated for each specific field or at least for soils of similar properties (Adamchuk et al. 2007, Olfs et al. 2012, Schirrmann et al. 2011). The derivation of specific models can be done with a selection of calibration samples; Adamchuk et al. (2011) proposed a targeted sampling strategy for on-the-go soil sensor data calibration.

Furthermore, it is noted that, if soil pH is rather extreme (less than pH 4 or more than pH 7), the Sb electrode reading shows a larger offset compared to the standard laboratory pH measurement. This is attributed to a memory effect of the Sb electrode that results from the pH measurements done in rapid succession, which inhibit equilibration of the Sb electrodes in order to capture the full range of pH values (Schirrmann et al. 2011).

Also, several mechanical and electrical problems are still to be solved, e.g. clogging of the soil sampler may occur as well as noisy pH readings and sample carryover from insufficiently cleaned pH electrodes. Leaf litter and organic matter on top of an acidic soil's surface may form a thin more alkaline layer (1–2 cm thick), which may distort test results; litter and biological soil crusts, e.g., from soil algae (Belnap and Lange 2003), may substantially alter spectroscopic properties of a soil surface, and thus the results of VIS/NIR measurements. Hence, soil tillage before pH analysis is advisable and typically part of on-the-go systems (No 8 in Fig. 3.9).

3.4.3.2. Mapping strategies

The mapping strategy applied is a critical factor for effectively and correctly gathering spatial distribution of soil properties. Covering this topic in full depth would go beyond the scope of this chapter. However, some fundamental aspects are briefly presented below.

Fig. 3.10 shows different standard and alternative soil sampling schemes (Schirrmann et al. 2011). Grid sampling with bulking (Fig. 3.10A) is the most commonly used strategy, because it combines samples from several subsample spots into one composite sample. This sampling method is frequently recommended by authorities such as the German Advisory Board for Agricultural Analysis (Boysen et al. 2000), or the British Ministry of Agriculture Fisheries and Food (2000). Lauzon et al. (2005) concluded that a grid spacing of 30 m or less is required to adequately assess the spatial variation of soil pH. Sampling at that grid size requires approximately 11 times as many soil samples as the commonly used 10 000 m² grid.

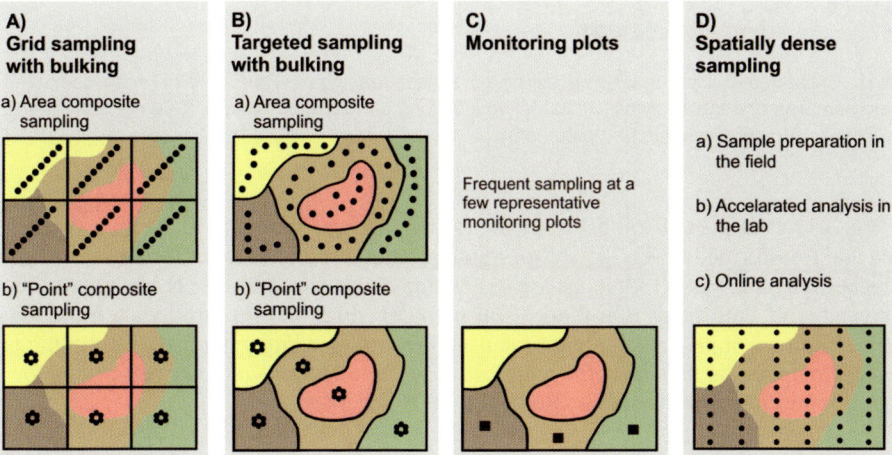

Fig. 3.10. Standard and alternative sampling strategies for soil pH mapping (Source: Schirrmann et al. 2011). Reprint with permission from the authors).

Targeted sampling with bulking (Fig. 3.10B) as well as monitoring plots (Fig. 3.10C) make use of previous intensive soil surveys or other auxiliary data to subdivide the field into irregular shaped zones, which are representative of classes of property characteristics (Griffin 1999, Mallarino and Wittry 2004, Schnug et al. 1994, Stamper et al. 2014). Monitoring plots (Fig. 3.10C) are defined, based on the assumption that a few sampling locations represent all the different sections of a site. However, data from such a typically small number of sampling locations may not reflect the full variability within a field.

Spatially dense sampling (Fig. 3.10D) is able to detect most of the small-scale variation in soil pH, and allows deriving reliable, high-resolution soil pH maps (Bianchini and Mallarino 2002, Bramley and Janik 2005, Brouder et al. 2005, Lauzon et al. 2005). For the spatially dense sampling, large numbers of samples are taken from various positions and with high density pattern. This sampling strategy complies best with the operating principle of on-the-go systems.

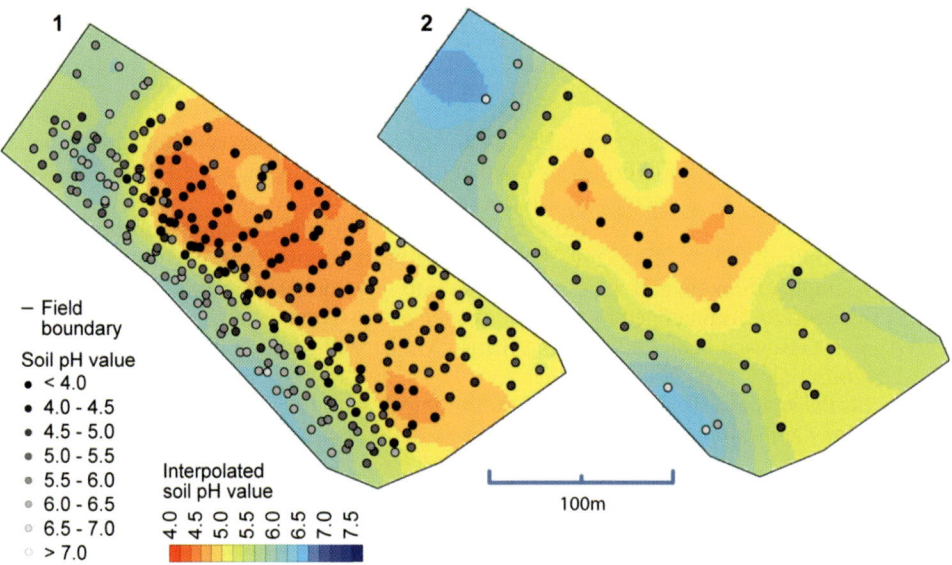

Fig. 3.11. Calibrated soil pH map from on-the-go sensor readings (1) and soil pH map computed from standard sampling and lab analysis (2); sampling locations are depicted as dots (Source: Schirrmann et al. 2011). Reprint with permission from the authors).

In Fig. 3.11 examples of soil pH maps are shown that were obtained (1) from computed, calibrated and transformed on-the-go sensor readings and (2) from standard sampling with subsequent laboratory analysis. Both maps are largely matching, the low soil pH zone is well defined in the centres of both maps. On-the-go soil mapping (Fig. 3.11-1) provides a more detailed, higher resolution soil pH map, simply because it is based on a larger number of sampling points.

Soil pH mapping is thus a useful preparatory work to identify representative sites for long-term pH monitoring projects (see section 3.4.2 of this chapter).

3.5. References

Adamchuk, V.I., Lund, E.D., Reed, T.M., Ferguson, R.B., 2007. Evaluation of an on-the-go technology for soil pH mapping. Precision Agriculture 8: 139–149.
Adamchuk, V.I., Morgan, M.T., Ess, D.R., 1999. An automated sampling system for measuring soil pH. 42.
Adamchuk, V.I., Viscarra Rossel, R.A., Marx, D.B., Samal, A.K., 2011. Using targeted sampling to process multivariate soil sensing data. Geoderma 163: 63–73.
ad-hoc AG Boden, 2005. Bodenkundliche Kartieranleitung. 5 ed., E. Schweizerbart'sche Verlagsbuchhandlung, Hannover, Stuttgart.
Ahern, C.R., Baker, D.E., Aitken, R.L., 1995. Models for relating pH measurements in water and calcium chloride for a wide range of pH, soil types and depths. Plant and Soil 171: 47–52.
Aitken, R., Moody, P., 1991. Interrelations between soil pH measurements in various electrolytes and soil solution pH in acidic soils. Soil Research 29: 483–491.
Al-Busaidi, A., Cookson, P., 2002. Methods of pH determination in calcareous soils of Oman: The effect of electrolyte and soil/solution ratio. Arab Gulf Journal of Scientific Research 20: 10–17.
Al-Busaidi, A., Cookson, P., Yamamoto, T., 2005. Methods of pH determination in calcareous soils: Use of electrolytes and suspension effect. Australian Journal of Soil Research 43: 541–545.
Alewell, C., 2003. Acid inputs into soils from acid rain. In: Rengel, Z. (Ed.), Handbook of Soil Acidity. Marcel Dekker, New York.
Bachmann, J., 1997. Wärmefluß und Wärmehaushalt. In: Blume, H.-P. et al. (Eds.), Handbuch der Bodenkunde. Ecomed, Landsberg/Lech, pp. 1–40.
Bates, R.G., 1973. Determination of pH; Theory and Practice. Wiley, New York.
Baucke, F.G.K., 1994a. The modern understanding of the glass electrode response. Fresenius' Journal of Analytical Chemistry 349: 582–596.
Baucke, F.G.K., 1994b. Thermodynamic origin of the sub-Nernstian response of glass electrodes. Analytical Chemistry 66: 4519–4524.
Belnap, J., Lange, O.L., 2003. Biological Soil Crusts: Structure, Function, and Management. Springer, Heidelberg.
Bianchini, A.A., Mallarino, A.P., 2002. Soil-sampling alternatives and variable-rate liming for a soybean-corn rotation. Agronomy Journal 94: 1355–1366.
Blake, L., 2005. Acid rain and soil acidification. In: Hillel, D. (Ed.), Encyclopedia of Soils in the Environment. Elsevier, Oxford, pp. 1–11.
Blossfeld, S., Gansert, D., 2007. A novel non-invasive optical method for quantitative visualization of pH dynamics in the rhizosphere of plants. Plant Cell and Environment 30: 176–186.
Blossfeld, S., Gansert, D., Thiele, B., Kuhn, A.J., Lösch, R., 2011. The dynamics of oxygen concentration, pH value, and organic acids in the rhizosphere of *Juncus* spp. Soil Biology and Biochemistry 43: 1186–1197.
Blume, H.P., Stahr, K., Leinweber, P., 2011. Bodenkundliches Praktikum. 3 ed. Spektrum Akademischer Verlag, Heidelberg.
Bolan, N.S., VCurtin, D., Adriano, D.C., 2004. Acidity. In: Hillel, D. (Ed.), Encyclopedia of Soils in the Environment. Academic Press, Burlington, MA, pp. 11–17.
Boldt, F.M., Heinze, J., Diez, M., Petersen, J., Borsch, M., 2004. Real-time pH microscopy down to the molecular level by combined scanning electrochemical microscopy/single-molecule fluorescence spectroscopy. Analytical Chemistry 76: 3473–3481.
Borchert, A., Olfs, H.-W., Pralle, H., Kohlbrecher, M., Trautz, D., 2011. Comparison of variable liming strategies in organic farming systems using online pH-measurements. In: Neuhoff, D. et al. (Eds.), Organic is Life – Knowledge for Tomorrow. Volume 1 – Organic Crop Production. Proceedings of the Third Scientific Conference of the International Society of Organic Agriculture Research (ISOFAR), held at the 17th IFOAM Organic World Congress in cooperation with the International Federation of Organic Agriculture Movements (IFOAM) and the Korean Organizing Committee (KOC), 28. September – 1. October 2011 in Namyangju, Korea Republic, pp. 21–24.
Boysen, P. et al., 2000. VDLUFA Standpunkt: Georeferenzierte Bodenprobenahme auf landwirtschaftlichen Flächen als Grundlage für eine teilflächenspezifische Düngung mit Grundnährstoffen. VDLUFA-Verlag, Darmstadt, Germany.
Bramley, R.G.V., Janik, L.J., 2005. Precision agriculture demands a new approach to soil and plant sam-

pling and analysis – Examples from Australia. Communications in Soil Science and Plant Analysis 36: 9–22.

Brouder, S.M., Hofmann, B.S., Morris, D.K., 2005. Mapping soil pH: Accuracy of common soil sampling strategies and estimation techniques. Soil Science Society of America Journal 69: 427–442.

Brown, D.J., Shepherd, K.D., Walsh, M.G., Dewayne Mays, M., Reinsch, T.G., 2006. Global soil characterization with VNIR diffuse reflectance spectroscopy. Geoderma 132: 273–290.

Bühler, H., Bucher, R., 1982. pH-Messung und Temperaturkompensation. GIT Fachzeitschrift für das Laboratorium 26: 839–844.

Calmano, W., without date. Determination of pH value (in German). Internet source, Institut für Umwelttechnik und Energiewirtschaft – IUE, TU Hamburg-Harburg, Germany.

Chang, C.W., Laird, D.A., Mausbach, M.J., Hurburgh C.R, Jr., 2001. Near-infrared reflectance spectroscopy – Principal components regression analyses of soil properties. Soil Science Society of America Journal 65: 480–490.

Conkling, B.L., Blanchar, R.W., 1988. A comparison of pH measurements using the antimony microelectrode and glass electrode. Agronomy Journal 80: 275–278.

Covington, A.K., Bütikofer, H.P., Camoes, M.F.G.F.C., Ferra, M.I.A., Rebelo, M.J.F., 1985. Procedures for testing pH responsive glass electrodes at 25, 37, 65 and 85 °C and determination of alkaline errors up to 1 mol dm^{-3} Na$^+$, K$^+$, Li$^+$. Pure and Applied Chemistry 57: 887–898.

Davis, L.E., 1943. Measurements of pH with the glass electrode as affected by soil moisture. Soil Science 56: 405–422.

Decker, M., Bause, S., Teichmann, P., Schneider, M., Vonau, W., 2017. Development of an automatic system for the on-site pH measurement of soil samples. Technisches Messen 84: 659–671.

Degner, R., 2012. pH-Messung: Der Leitfaden für Praktiker. Wiley-VCH, Weinheim.

DIN, 2000. DIN 19261, pH-Messung, Messverfahren mit Verwendung potentiometrischer Zellen. Beuth-Verlag, Berlin.

DIN, 2005. Soil quality – Determination of pH; DIN ISO 10390:2005-12. Beuth, Berlin.

DIN, 2012. DIN 19260:2012-10. pH-Messung – Allgemeine Begriffe. Beuth, Berlin.

Elberling, B., Matthiesen, H., 2007. Methodologically controlled variations in laboratory and field pH measurements in waterlogged soils. European Journal of Soil Science 58: 207–214.

Faget, M., Blossfeld, S., Von Gillhaußen, P., Schurr, U., Temperton, V., 2013. Disentangling who is who during rhizosphere acidification in root interactions: combining fluorescence with optode techniques. Frontiers in Plant Science 4.

Fieldes, M., Perrott, K., 1966. The nature of allophane soils: 3. Rapid field and laboratory test for allophane. New Zealand Journal of Science 9: 623–629.

Galster, H., 1990. pH-Messung: Grundlagen, Methoden, Anwendungen, Geräte. Wiley-VCH, Weinheim.

Gascho, G.J., Parker, M.B., Gaines, T.P., 1996. Reevaluation of suspension solutions for soil pH. Communications in Soil Science and Plant Analysis 27: 773–782.

Gavriloaiei, T., 2012. The influence of electrolyte solutions on soil pH measurements. Revista de Chimie 63: 396–400.

Griffin, S.J., 1999. Directed soil sampling as a means of increasing nutrient map accuracy using complementary precision farming data. In: Stafford, J.V. (Ed.), Precision Agriculture '99: Proceedings of the 2nd European Conference on Precision Agriculture Part I. Sheffield Academic, Sheffield, UK, pp. 141–149.

Grybos, M., Davranche, M., Gruau, G., Petitjean, P., 2007. Is trace metal release in wetland soils controlled by organic matter mobility or Fe-oxyhydroxides reduction? Journal of Colloid and Interface Science 314: 490–501.

Hawkes, C.V., DeAngelis, K.M., Firestone, M.K., 2007. CHAPTER 1 – Root Interactions with Soil Microbial Communities and Processes A2 – Cardon, Zoe G. In: Whitbeck, J.L. (Ed.), The Rhizosphere. Academic Press, Burlington, pp. 1–29.

Hendershot, W.H., Lalande, H., Duquette, M., 2006. Soil Reaction and Exchangeable Acidity. In: Carter, M.R., Gregorich, E.G. (Eds.), Soil Sampling and Methods of Analysis. Taylor & Francis, Boca Raton, USA, pp. 173–178.

Henderson, B.L., Bui, E.N., 2002. An improved calibration curve between soil pH measured in water and CaCl. Australian Journal of Soil Research 40: 1399–1405.

IUSS Working Group WRB, 2014. World reference base for soil resources 2014. World Soil Resources

Reports No. 106. FAO, Rome.

Jahn, R. et al., 2006. Guidelines for Soil Description. 4 ed. Food and Agriculture Organization of the United Nations, Rome.

Jaillard, B., Plassard, C., Hinsinger, P., 2003. Measurements of H^+ fluxes and concentrations in the rhizosphere. In: Rengel, Z. (Ed.), Handbook of Soil Acidity. Marcel Dekker, New York.

Kahlert, H., Steinhardt, T., Behnert, J., Scholz, F., 2004. A new calibration free pH-probe for in situ measurements of soil pH. Electroanalysis 16: 2058–2064.

Kheiralla, A.F., El-Fatih, W.T., Abdellatief, M.K., El-Talib, Z.M., 2016. Design and development of on-the-go soil pH mapping system for precision agriculture, Proceedings of 2016 Conference of Basic Sciences and Engineering Studies, SGCAC 2016, pp. 192–195.

Lahl, K., Unger, C., Emmerling, C., Broer, I., Thiele-Bruhn, S., 2012. Response of soil microorganisms and enzyme activities on the decomposition of transgenic cyanophycin-producing potatoes during overwintering in soil. European Journal of Soil Biology 53: 1–10.

Lauzon, J.D., O'Halloran, I.P., Fallow, D.J., Von Bertoldi, A.P., Aspinall, D., 2005. Spatial variability of soil test phosphorus, potassium, and pH of Ontario soils. Agronomy Journal, 97: 524–532.

Libohova, Z. et al., 2014. Converting pH1:1 H2O and 1:2CaCl2 to 1:5 H2O to contribute to a harmonized global soil database. Geoderma 213: 544–550.

Little, I., 1992. The relationship between soil pH measurements in calcium chloride and water suspensions. Soil Research 30: 587–592.

Lund, E.D., Adamchuk, V.I., Collings, K.L., Drummond, P.E., Christy, C.D., 2005. Development of soil pH and lime requirement maps using on-the-go soil sensors, Precision Agriculture 2005, ECPA 2005, pp. 457–464.

Mallarino, A.P., Wittry, D.J., 2004. Efficacy of grid and zone soil sampling approaches for site-specific assessment of phosphorus, potassium, pH, and organic matter. Precision Agriculture 5: 131–144.

Matthiesen, H., 2004. In situ measurement of soil pH. Journal of Archaeological Science 31: 1373–1381.

McBratney, A.B., Pringle, M., 1997. Spatial variability in soil implications for precision agriculture. In: Stafford, J.V. (Ed.), Precision Agriculture. BIOS, Oxford, UK, pp. 3–31.

McCarty, G.W., Reeves III, J.B., 2006. Comparison of near infrared and mid infrared diffuse reflectance spectroscopy for field-scale measurement of soil fertility parameters. Soil Science 171: 94–102.

Mettler-Toledo, 2007. Anleitung zur Messung von pH – Theorie und Praxis von pH-Anwendungen im Labor. Mettler-Toledo AG, Analytical, Schwerzenbach, Schweiz.

Miller, R.O., Kissel, D.E., 2010. Comparison of soil pH methods on soils of North America. Soil Science Society of America Journal 74: 310–316.

Minasny, B., McBratney, A.B., Brough, D.M., Jacquier, D., 2011. Models relating soil pH measurements in water and calcium chloride that incorporate electrolyte concentration. European Journal of Soil Science 62: 728–732.

Ministry of Agriculture Fisheries and Food, 2000. Fertiliser : recommendations for agricultural and horticultural crops (RB209). 7 ed. Stationery Office, London, UK.

Motellier, S., Noiré, M.H., Pitsch, H., Duréault, P., 1995. pH determination of clay interstitial water using a fiber-optic sensor. Sensors and Actuators: B. Chemical 29: 345–352.

Muñoz, K. et al., 2017. Physicochemical and microbial soil quality indicators as affected by the agricultural management system in strawberry cultivation using straw or black polyethylene mulching. Applied Soil Ecology 113: 36–44.

Olfs, H.-W., Borchert, T., Trautz, D., 2012. Soil pH maps derived from on-the-go pH-measurements as basis for variable lime application under German conditions: Concept development and evaluation in field trials. Proceedings of the 11th Int. Conf. on Precision Agriculture, Indianapolis, USA, 1–10.

Oman, S., 2000a. A step to a uniform definition and interpretation of the suspension effect. Talanta 51: 21–31.

Oman, S., Camões, I., Powell, K.J., Rajagopalan, R., Spitzer, P., 2007. Guidelines for potentiometric measurements in suspensions Part B. Guidelines for practical pH measurements in soil suspension (IUPAC Recommendations 2006). Pure Appl. Chem. 79: 81–86.

Oman, S.F., 2000b. On the seventieth anniversary of the "suspension effect": A review of its investigations and interpretations. Acta Chimica Slovenica 47: 519–534.

Oman, S.F., 2012. A conventional method for valid "actual soil pH" measurement. Acta Chimica Slovenica 59: 969–973.

Overbeek, J.T.G., 1953. Donnan-e.m.f. and suspension effect. Journal of Colloid Science 8, 593–605.
Pallmann, H., 1930. Die Wasserstoffaktivität in Dispersionen und kolloid-dispersen Systemen. Kolloidchem. B. 30: 34–405.
Parker, D.R., 2005. Aluminum speciation. In: Hillel, D. (Ed.), Encyclopedia of Soils in the Environment. Elsevier, Oxford, pp. 50–56.
Paul, E.A., 2007. Soil Microbiology, Ecology and Biochemistry. 3rd ed. Academic Press, Burlington, MA.
Pirie, A., Singh, B., Islam, K., 2005. Ultra-violet, visible, near-infrared, and mid-infrared diffuse reflectance spectroscopic techniques to predict several soil properties. Australian Journal of Soil Research 43: 713–721.
Rengel, Z., 2003. Handbook of Soil Acidity. CRC Press.
Schaller, G., Fischer, W.R., 1981. Die Verwendung von Antimon-Elektroden zur pH-Messung in Böden. Z. Pflanzenernähr. Bodenk. 144: 197–204.
Schimming, C.-G., 2011. Säurebelastung. In: Blume, H.-P., Horn, R., Thiele-Bruhn, S. (Eds.), Handbuch des Bodenschutzes. Wiley-VCH, Weinheim, pp. 270–286.
Schirrmann, M., Gebbers, R., Kramer, E., 2013. Performance of automated near-infrared reflectance spectrometry for continuous in situ mapping of soil fertility at field scale. Vadose Zone Journal 12.
Schirrmann, M., Gebbers, R., Kramer, E., Seidel, J., 2011. Soil pH mapping with an on-the-go sensor. Sensors 11: 573–598.
Schnug, E., Haneklaus, S., Murphy, D.P. 1994. Equifertiles – an innovative concept for efficient sampling in the local resource management of agricultural soils. AAB, Warwick, pp. 63–72.
Schreiber, C.M. et al., 2012. Monitoring rhizospheric pH, oxygen, and organic acid dynamics in two short-time flooded plant species. Journal of Plant Nutrition and Soil Science 175: 761–768.
Schröder, C.R., Polerecky, L., Klimant, I., 2007. Time-resolved pH/pO(2) mapping with luminescent hybrid sensors. Analytical Chemistry 79: 60–70.
Schwabe, K., 1976. pH-Messtechnik. 4 ed. Verlag Theodor Steinkopff, Dresden.
Sethuramasamyraja, B. et al., 2008. Agitated soil measurement method for integrated on-the-go mapping of soil pH, potassium and nitrate contents. Computers and Electronics in Agriculture 60: 212–225.
Soil Survey Staff, 2004. Soil Survey Laboratory Information Manual. Soil Survey Investigations Report No. 45, Version 2.0. U.S. Department of Agriculture, Natural Resources Conservation Service, Lincoln, USA.
Soil Survey Staff, 2010. Keys to Soil Taxonomy. 11 ed. United States Department of Agriculture, Natural Resources Conservation Service, Lincoln, USA.
Soriano-Disla, J.M., Janik, L.J., Viscarra Rossel, R.A., MacDonald, L.M., McLaughlin, M.J., 2014. The performance of visible, near-, and mid-infrared reflectance spectroscopy for prediction of soil physical, chemical, and biological properties. Applied Spectroscopy Reviews 49: 139–186.
Stahl, H. et al., 2006. Time-resolved pH imaging in marine sediments with a luminescent planar optode. Limnology and Oceanography-Methods 4: 336–345.
Stamper, D.J., Agouridis, C.T., Edwards, D.R., Purschwitz, M.A., 2014. Effect of soil sampling density and landscape features on soil test phosphorus. Applied Engineering in Agriculture 30: 773–781.
Suarez, D.L., 1998. Thermodynamics of the soil solution. In: Sparks, D.L. (Ed.), Soil Physical Chemistry. CRC Press, Boca Raton, USA, pp. 97–134.
Tamogami, J., Kikukawa, T., Miyauchi, S., Muneyuki, E., Kamo, N., 2009. A tin oxide transparent electrode provides the means for rapid time-resolved pH measurements: application to photoinduced proton transfer of bacteriorhodopsin and proteorhodopsin. Photochemistry and Photobiology 85: 578–589.
Tan, K.H., 2011. Principles of Soil Chemistry. 4th ed. CRC Press, Boca Raton, FL.
Thiele-Bruhn, S., Wessel-Bothe, S., Aust, M.O., 2015. Time-resolved in-situ pH measurement in differently treated, saturated and unsaturated soils. Journal of Plant Nutrition and Soil Science 178: 425–432.
Tiunov, A.V., Scheu, S., 1999. Microbial respiration, biomass, biovolume and nutrient status in burrow walls of Lumbricus terrestris L. (Lumbricidae). Soil Biology and Biochemistry 31: 2039–2048.
Viscarra Rossel, R.A. et al., 2005. Field measurements of soil pH and lime requirement using an on-the-go soil pH and lime requirement measurement system, Precision Agriculture 2005, ECPA 2005, pp. 511–520.
Viscarra Rossel, R.A., McBratney, A.B., 1997. Preliminary experiments towards the evaluation of a suitable soil sensor for continuous, 'on-the-go' field pH measurements. In: Stafford, J.V. (Ed.), Precision Agriculture. BIOS, Oxford, UK, pp. 493–502.

Viscarra Rossel, R.A., Walter, C., 2004. Rapid, quantitative and spatial field measurements of soil pH using an ion sensitive field effect transistor. Geoderma 119: 9–20.

Viscarra Rossel, R.A., Walvoort, D.J.J., McBratney, A.B., Janik, L.J., Skjemstad, J.O., 2006. Visible, near infrared, mid infrared or combined diffuse reflectance spectroscopy for simultaneous assessment of various soil properties. Geoderma 131: 59–75.

Vohland, M., Harbich, M., Ludwig, M., Emmerling, C., Thiele-Bruhn, S., 2016. Quantification of soil variables in a heterogeneous soil region with VIS-NIR-SWIR data using different statistical sampling and modeling strategies. IEEE Journal of Selected Topics in Applied Earth Observations and Remote Sensing, PP: 1–11.

Wang, Y. et al., 2015. Soil pH value, organic matter and macronutrients contents prediction using optical diffuse reflectance spectroscopy. Computers and Electronics in Agriculture 111: 69–77.

Soil water parameters

4. Soil water content measurement

Heye R. Bogena and Lutz Weihermüller

4.1. Introduction

Soil water content (often denoted as soil moisture) is an important state variable in the terrestrial system because it controls the exchange of water and energy between the land surface and the atmosphere. Soil water content is highly variable in space and time with characteristic length scales ranging from a few centimeters up to several kilometers and characteristic time scales ranging from minutes up to years (Vereecken et al. 2014). Information on soil water content dynamics is also important for optimizing agricultural management and improving our understanding of biogeochemical (Zhu et al. 2013), vadose zone (Vereecken et al. 2008), hydrological (Lee et al. 2011), and atmospheric processes (Seneviratne et al. 2010). Soil water content data has been used in many studies to analyze spatial variability of soil water content at a range of scales, including the field (Bell et al. 1980, Famiglietti et al. 1998), catchment (Rosenbaum et al. 2012, Western et al. 2004), regional (Romshoo 2004, Zhao et al. 2013), and continental scale (Entin et al. 2000, Li and Rodell 2013).

Since the first attempts to characterize the variability of soil water content at the field scale (Nielsen et al. 1973), tremendous progress has been made with respect to measuring soil water content. In the following, we review the most commonly used techniques to measure soil water content at the field scale, including point measurements using destructive sampling, electromagnetic soil water content sensors, and hydrogeophysical methods as emerging technologies to monitor field-scale or even field average soil water contents.

Generally speaking, soil water content can only be measured directly by destructive sampling and all sensor based soil water content 'measurements' are indirect methods, where a (geo-)physical quantity will be related to soil water content.

Soil water contents can be provided as gravimetric soil water content θ_G (g g^{-1}, or more precisely g_{water} per g_{soil}), which is often also provided in relative terms (%). The other possibility is the expression as volumetric fraction θ_v (m^3 m^{-3}, or more precise m$^3_{water}$ per m$^3_{soil}$). Even if g g^{-1} and m^3 m^{-3} are unit-less (as the units cancel), the units should be always reported to indicate whether gravimetric or volumetric soil water content is meant.

In the following, we introduce different soil water content sensor types with their advantages and disadvantages and provide a general overview of the sensing techniques in use at the end of the chapter.

4.2. Soil water content sensor types – Method selection

4.2.1. Destructive soil water content measurement

As mentioned in the introduction, soil water content can be only measured directly by destructive sampling. The easiest method is the sampling of disturbed probes in the field from different vertical or horizontal locations. Therefore, a certain amount of field wet soil (~100 g) will be taken and transferred to the laboratory (see Fig. 4.1). In order to determine the volumetric water content the volume of the sampled soil must be known. To this end, rings of known volume (so-called Kopecky-rings) are pushed gently into the soil and dug out afterwards (see Fig. 4.1). All soil overlying the ring is then cut off. Either the entire ring (containing the soil) will be transferred to the laboratory for further analysis or the extracted soil volume is transferred to air tight containers or plastic bags. The second possibility is preferable whereever large numbers of samples must be taken and/or only few Kopecky-rings are available. Classical Kopecky-rings have volumes of 100 to 250 cm^3.

For both methods the soil should be stored between sampling and laboratory analysis in air tight containers, plastic bags, or in closed Kopecky-rings (see Fig. 4.1) to rule out changes of the water content over time. Between sampling and analysis the probes should be stored in a cool environment (e.g., at 4 °C). After transfer to the lab, the weight of the field wet sample must be determined by weighing and the sample is dried at 105 °C for 48 h. After drying and cooling down (cooling should take place in an desiccator to avoid uptake of water from the atmosphere) the dry weight is determined. By knowing of wet and dry weights the water loss is determined and either gravimetric or volumetric water content is calculated from the difference in weight and the volume.

Fig. 4.1. Undisturbed soil sampling for the measurement of volumetric water content using 100 cm^3 Kopecky-rings (upper panel) and box with sampled Kopecky-rings and additional disturbed samples for gravimetric water content determination (Fotos: Forschungszentrum Jülich GmbH).

4.2.2. Electromagnetic (EM) sensors

The volumetric soil water content, θ_v (m³ m⁻³), can be estimated using electromagnetic (EM) sensors. To this end, the measured EM sensor response in the medium under consideration is related to soil water content by a petrophysical relationship. Petrophysical relationships are functions relating a measured physical quantity (e.g., relative dielectric permittivity of a soil) with the volumetric water content. One of the most commonly used petrophysical relationship to relate the relative dielectric permittivity to volumetric water content is the empirical function of Topp et al. (1980). The strong dependence of the EM signal on soil water stems from the high relative dielectric permittivity of free water ($\varepsilon \approx 80$) compared to other soil constituents like minerals ($\varepsilon \approx 1$–3) or air ($\varepsilon = 1$). In general, the relative dielectric permittivity is a measure of how strong an electromagnetic field interacts with the dielectric medium such as water or soil. Due to their dipole characteristics, water molecules are polarized more easily in response to an electromagnetic field than other solid soil constitutes or soil air. Therefore, in a wet soil a greater electromagnetic flux is generated per unit charge emitted by an EM sensor compared to that induced in a dry soil or the electromagnetic wave propagation is quicker in dry soils compared to wet soils.

Electromagnetic estimates of soil water content have gained popularity as technology has improved and sensors became cheaper in price. These sensors, including time domain reflectometry (TDR), time domain transmission (TDT), and capacitance devices, are valuable field and laboratory techniques for monitoring water and in some cases also bulk electrical conductivity. We will discuss the basic technology of these sensor types in greater detail below.

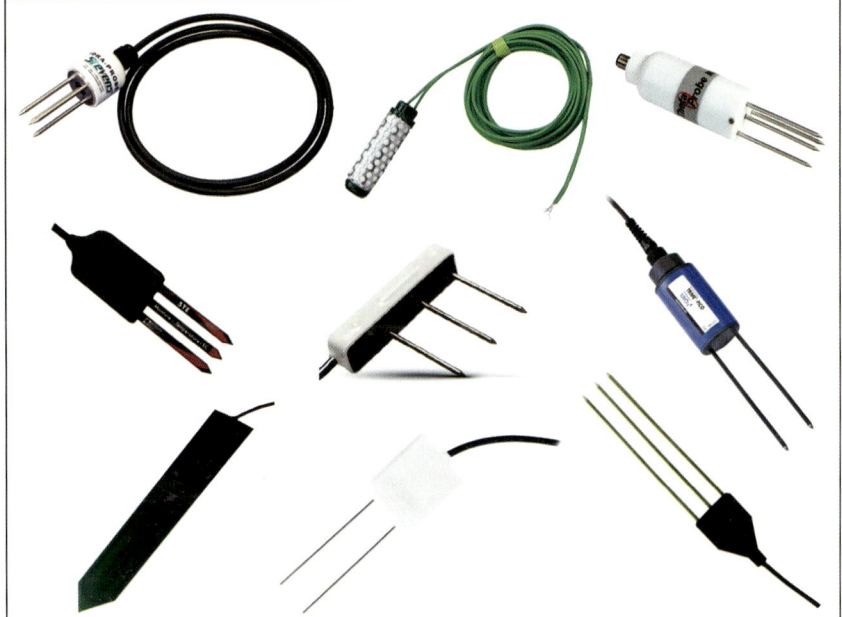

Fig. 4.2. Selected soil water content sensors. From upper left to lower right: Hydra-Probe (Stevens Water Monitoring Systems, Inc.), Watermark Sensor (IRROMETER), ThetaProbe (Delta-T Devices Inc.), 5TE (Decagon Devices, Inc.), GS3 (METER Group), TRIME-PICO (IMKO Micromodultechnik GmbH), SMT100 (TRUEBNER GmbH), CS655 Sensor (Campbell Scientific, Inc.), and CS635 TDR-probe (Campbell Scientific, Inc.). Note, that the sensors are not shown in same scale.

4.2.3. Time domain reflectometry

Time domain reflectometry (TDR) was first introduced to soil science by Topp et al. (1980) and since then has developed into a standard method for measuring soil water content. The relative dielectric permittivity (ε) in the soil is determined from the velocity of an electromagnetic wave emitted by a pulse generator ('cable tester') and passed along the waveguides of the TDR probe. The propagation velocity is determined from the measured travel time along a TDR probe of known length. For more information on the principles of TDR the reader is referred to the comprehensive review of Robinson et al. (2003). Several TDR probes can be combined in a network configuration using multiplexing systems (Heimovaara and Bouten 1990, Weihermüller et al. 2013). However, such networks of probes are still restricted to local applications (e.g., trenches or small field plots) because of limitations on cable length (<20 m) for accurate TDR measurements. Recent developments embedded the cable tester in the sensor head to overcome the problem of restricted cable length and the sensors can be also connected to a wide range of commercially available data loggers because only the measured analog/digital measurement signal will be transferred and stored. At larger scales, many studies of field-scale variability of soil water content have used TDR for manual sampling campaigns. For instance, a detailed investigation of soil water content pattern dynamics and connectivity in spatial patterns has been undertaken at the Tarrawarra catchment (Australia), which involved the spatial characterization of soil water content in the top 0.30 m at 520 sampling locations (Grayson et al. 1997, Western et al. 2001).

4.2.4. Capacitance sensors

In the last two decades, frequency domain (FD) sensors have emerged as cost-effective alternatives to TDR sensors. FD sensors consist of an oscillating circuit, typically located in the sensor head, and a sensing part (e.g., sensor prongs) embedded in the soil. There are two types of FD sensors: Capacitance sensors and electrical impedance (EI) sensors.

Capacitance sensors are relatively inexpensive and easy to operate. The principle of the capacitance method is to incorporate the soil medium that surrounds the sensor prong as part of the dielectric of the sensor capacitor (Bogena et al. 2007). The relative dielectric permittivity of the soil is then determined by measuring the charge time from a starting voltage to a voltage with an applied capacitor voltage. A popular example of this sensor type is the family of capacitance sensors by METER Group (e.g., EC-5, GS3, 5TE).

EI sensors are comprised of a sinusoidal oscillator, a fixed impedance coaxial transmission line, and sensor probes which are buried in the soil. The oscillator signal is propagated along the transmission line into the sensor probe. If the impedance of the sensor probe differs from that of the transmission line, a proportion of the incident signal is reflected back along the line towards the signal source (Gaskin & Miller 1996). The reflected component interferes with the incident signal causing a voltage standing wave to be set up on the transmission line is a function of soil dielectric permittivity and thus can be related to soil water content. Popular examples of this sensor type are the ML3 ThetaProbe Soil Moisture Sensor (Delta-T Devices, Inc.) and the Hydra Probe Soil Sensor (Stevens® Water Monitoring System, Inc.).

Typically, FD sensors operate at a measurement frequency between 50 to 100 MHz. Therefore, sensor output is to some extent influenced by the electrical conductivity and imaginary part of the dielectric permittivity of the soil (Kelleners et al. 2005, Kizito et al. 2008, Robinson et al. 2005), which affects the accuracy of soil water content measurements with capacitance sensors. Because of its low cost, this sensor type it has become very attractive for sensor network applications (e.g., Rosenbaum et al. 2012) (see section 4.2.6.2).

4.2.5. Time domain transmission sensors

Another cost-effective electromagnetic sensor type is the time domain transmission (TDT) sensor, which measures the propagation velocity of an electromagnetic wave along a closed transmission line. Currently, there are different TDT approaches available. For instance, Blonquist et al. (2005) reported on the Acclima TDT sensor (McCready et al. 2009), which uses a waveform interpretation process similar to conventional TDR systems. According to Blonquist et al. (2005), the Acclima TDT sensor has a very similar measurement accuracy compared to reference TDR systems (within ±3 permittivity units in the permittivity range of 9 to 80). Unfortunately, the Acclima TDT sensor does not allow direct insertion in natural soils and thus needs to be buried. Other TDT sensors use the oscillation frequency of a ring oscillator on a printed circuit board to approximate a propagation velocity, and such designs can easily be inserted in undisturbed soils. Because these TDT sensors operate at higher frequencies than capacitance sensors, they are expected to provide a higher measurement accuracy (Blonquist et al. 2005). One example of such a TDT sensor is the SPADE sensor (sceme.de GmbH i.G., Horn-Bad Meinberg, Germany; Hübner et al. 2009), which was recently tested by Qu et al. (2013). The SPADE sensor generates a pulse, which is inverted and then fed back to the input of the line driver resulting in an "oscillation" frequency that mainly depends on the dielectric permittivity of the surrounding medium (between 150 MHz in water and 340 MHz in air). Qu et al. (2013) found that the SPADE sensor showed good agreement with TDR measurements after consideration of temperature effects on the sensor readings. Recently, the SMT100 sensor (TRUEBNER GmbH, Germany) was introduced, which can be seen as a successor of the SPADE sensor as it uses the same measurement principle. According to Bogena et al. (2017) the accuracy of the SMT100 is about 1 vol.% under ideal conditions (i.e., using liquid dielectric media).

4.2.6. Method selection with respect to research objective

4.2.6.1. Continuous data recording or single observations

Depending on the operation mode (continuous or campaign style) different types of soil water content sensor must be used. Where continuous data recording is required, the sensors are installed once in the soil profile and then stay there untouched for a long period of time. Typically, specific pre-installation tools are used prior to sensor installation to prevent the sensor rods being destroying during insertion into the soil. Therefore, the sensor rods do not have to be overly robust and may be simply constructed. Therefore, low-cost sensor designs, e.g., printed circuit boards, can be used. In contrast, if sensors have to be used for single observations in a campaign style, e.g., for the purpose of mapping soil water content pattern of an arable field, the sensor rods need to be specially ruggedized to allow repeated measurements without degrading the sensor rods. Because of the small scale variability of the soil water content, it is important to conduct several repeat measurements in each location (e.g., 3 to 5 repeats).

4.2.6.2. Soil water sensor networks

In recent years, wireless soil water content sensor networks to observe spatio-temporal soil water content variability in near real-time have emerged (e.g., Bogena et al. 2010). They allow bridging the gap between local and regional scale measurements (e.g., remote sensing) (Robinson et al. 2008). Wireless sensor networks typically consist of hundreds of soil moisture sensors that transmit information to a main server with wireless communication technology. One of the first low-cost mesh networking proprietary standards used for wireless sensor networks is ZigBee (Valente et al. 2006). However, the 2.4 GHz low-power radio module of ZigBee limits the communication range between sensor nodes (less than 1 km). The recently introduced LoRa system for long-range, low-power, low-bitrate, and wireless communication offers network coverage of several kilometers (Augustin et al. 2016). Because of the multitude of soil water content measurements within the sensor network, the interpretation of the sensor signal should be straightforward and unambiguous. In order to maximize the number of sensor nodes, the soil water content sensors must be as inexpensive as possible without compromising sensor accuracy too strongly. Therefore, capacitance and TDT sensors are currently considered to be most appropriate for use in wireless soil water content sensor networks (Bogena et al. 2017, Rosenbaum et al. 2010). Recent applications of wireless sensor networks include monitoring of soil water content combined with salinity in irrigated fields to optimize irrigation management (Yu et al. 2013), spatio-temporal observation of soil water content from the hillsope to the catchment scale (Rosenbaum et al. 2012, Qu et al. 2015, Martini et al. 2015), and validation of remote sensing data (Bircher et al. 2012, Rötzer et al. 2014).

4.2.6.3. On-site infrastructure and other measurements

The infrastructure needed for sensor based soil water content estimation greatly differs between sensor types and sensor applications. For single point measurements no infrastructure is generally required and most sensors allow the acquisition of soil water content by the use of handheld displays or loggers operating on batteries. For long-term measurements with TDT and capacitance sensors, small battery-powered logger devices exist which sometimes even have remote data transfer capabilities, whereby the loggers are often restricted to a fixed number of sensors that may be attached to them. Where a greater number of sensors is needed most TDT and capacitance sensors may also be attached to other commercially available logger systems.

Most TDR systems on the other hand require a specific setup consisting of a cable tester, multiplexers, and the sensors (antennas) (Weihermüller et al. 2013). This setup is often higher in complexity and requires a certain level of expert knowledge. Currently, some manufacturers tried to miniaturize the periphery (cable tester and multiplexers) and also have full-built-in solutions for easy applications.

The basic infrastructure of a wireless network consists of three different components: coordinator, router, and end device (Bogena et al. 2010). The coordinator unit is the top of the network tree storing information about the sensor network and providing a link to other networks. Each sensor network consists of a single coordinator and typically more than one router units acting as relay stations. Both coordinator and router units need electrical power supply (e.g., solar panel) and a mast for the antennas and thus are located on the surface, whereas the sensor units with the sensors can be fully buried in the ground.

4.2.6.4. Soil type: Depth of interest, soil texture, obstacles (roots, skeleton)

Soil water content sensors can be installed in the field at all depths, and their depth of installation is only restricted by accessibility. As a general rule, the sensors should not be installed too close to the soil surface because the EM soil water content sensors are sensitive to a certain soil volume.TDT and capacitance sensors measure a fairly small soil volume and can be installed closer to the surface, whereas TDR sensors should be at least 5 cm below the surface to avoid atmospheric influence in the sensor readings. Installation in skeleton rich soil or soils with larger rock fragments is critical because good contact between the sensor and the soil is necessary and rocks, larger roots, and skeleton might hamper installation or may even destroy the sensor itself. In such cases, ruggedized soil water content sensors with relatively short sensor probes should be used. Special installation devices are provided for some sensors to facilitate their installation in such environments.

Some EM soil water sensors such as FD and TDR might not work properly in saline soils because of the large loss of signal caused by the high electrical conductivity of such soils (Nichol et al. 2002).

High skeleton, rock, or root content in the soils may also result in unrealistic soil water content estimates because the petrophysical relationships used to translate the measured signal quantity to soil water content are not applicable to such environments. In organic rich soils (e.g., turf) special petrophysical relationships or calibration functions are also required, which are available in the literature (e.g., Nagare et al. 2011, Nemali et al. 2007, Vaz et al. 2013, Bircher et al. 2016).

4.3. Implementation of soil water content sensors

4.3.1. Installation of soil water content sensors

The installation of soil water content sensors is fairly easy and can be either performed from a borehole or from an access trench. In both cases good hydraulic contact between the soil and the sensor must be ensured and soil compaction in direct vicinity of the sensor should be avoided. Sensors should not be installed at the interface between soil horizons because the measurements will get information from both soil layers at differents ratios, which also depend on the actual soil water content status of the two layers as was shown by Rejiba et al. (2005) and Hinnel et al. (2006) for TDR probes.

4.3.1.1. Self-construction of TDR sensors

Because both capacitance and TDT sensors are complex designs self-construction of these sensors is not possible. TDR probes (this refers to the TDR antennas only) can be easily self-constructed, whereby the general geometry should follow the recommendations of Knight (1992).

4.3.1.2. Sensor calibration

As already mentioned, the EM sensors measure the apparent dielectric permittivity of a soil, which needs to be converted to soil water content. The relationship between apparent dielectric permittivity and soil water content depends on soil properties, e.g., soil texture, organic carbon content, bulk density, and soil structure. Usually, some general calibration equations are provided by manufacturers for both mineral and organic soils. However, to conduct more accurate measurements it is recommended to specifically calibrate the probes. For such a calibration undisturbed samples are taken from each soil horizon. If soil sensors are to be deployed in larger areas, e.g., as part of a wireless sensor network, the main soil types need to be sampled.

In a first step, these samples are fully water saturated. Next, the samples are dried out at room temperature and the volumetric soil water content and dielectric permittivity is determined at regular time intervals. The volumetric soil water content is determined from the weight of the sample, the known sample volume, and the dry weight of the sample determined at the end of the experiment by oven-drying (105 °C for mineral and 65 °C for organic soils for 48 h). If soil shrinkage occurs, e.g., due to high contents of clay or organic matter, either these samples should be discarded or the additional pore space from shrinkage needs to be subtracted from the total soil porosity.

The apparent dielectric permittivity of the soils determined from the sensor readings can be fitted to an empirical (e.g., Topp's) equation (Topp et al. 1980) or a semi-empirical model such as the complex refractive index model (CRIM) proposed by Birchak et al. (1974):

$$\theta = 100 * \frac{K_a^\beta - (1-\eta) \cdot K_s^\beta - \eta K_{air}^\beta}{K_w(T)^\beta - K_{air}^\beta} \qquad (4.1)$$

where, K_a is the measured sensor apparent dielectric permittivity, the shape factor ß is typically assumed to be 0.5 (Pepin et al. 1995), the dielectric permittivity of the solid phase K is typically assumed to be 4.4 (Robinson et al. 2004), the dielectric permittivity of air K_{air} is 1, the mean porosity, η, needs to be estimated and the temperature dependent dielectric permittivity of water, K_w, for 25 °C is 78.54 according to Weast et al. (1986) and θ is given in [vol. %].

4.3.1.3. Temperature effects on soil water content measurements

The effects of temperature on soil water content measurements with dielectric sensors are known to be complicated (e.g., Wraith and Or 1999). It is important to distinguish the effect of temperature on the dielectric measurement from the effect of temperature on the dielectric properties of the soil. Previous laboratory experiments have shown that temperature affects capacitance soil water content sensors due to the temperature sensitivity of the sensor electronics (Bogena et al. 2007), which can be corrected for temperatures between 5 and 40 °C (Rosenbaum et al. 2011). On the other hand, Bogena et al. (2017) showed that the electronics of the recently developed SMT100 sensor are not influenced by temperature effects. Thus in most soils, any temperature sensitivity of the measured soil moisture is related to the temperature dependency of the dielectric permittivity of water. Since the SMT100 sensor also measures temperature, users can easily correct for this temperature effect (see Bogena et al. 2017).

4.3.1. Sensor installation guide (required equipment, soil preparation, on-site installation steps)

Special attention needs to be paid to the installation of the soil water content sensor. The sensor should be inserted horizontally into the soil and it must be ensured that the sensor rods have good contact with the soil. The contact quality can be controlled by inserting at least two sensors at a distance of about 5 cm. If the soil water content readings of both sensors are within 5 vol.% of each other, it may be assumed that sufficient contact between sensor rods and soil matrix has been achieved. It has to be noted, that in most cases small air pockets collapse after some hours or days due to soil settlement. Sensors should not be installed directly one below the other to avoid 'shading effects' of the above top sensor on the sensor installed below. As mentioned earlier, some distance should be kept to the soil surface, whereby this distance depends on the sensor's measurement volume. Information about the sensor measurement volume is generally provided by the manufacturer.

4.3.2. Measurement and sensor maintenance

After installation the sensors do not require maintenance so that long data recordings are feasible without recalibration.

4.3.3. Raw data processing and measurement error handling

For most sensors either the raw signal (voltage, or travel time) can be stored, which must be translated to water content using appropriate petrophysical relationships such as the empirical function proposed by Topp et al. (1980) or physics based approaches such as the CRIM model (Eq. 4.1). Alternatively, calculated water contents may be stored. If raw data are stored, different petrophysical relationships may be easily applied to the data after the measurement. Especially, for TDR readings some errors may be detectable, which are caused by the algorithms used in the cable tester to determine the travel times of the EM wave. Errors associated with this algorithm can only be detected if the raw waveforms are stored and analyzed manually. Unfortunately, this procedure requires an advanced level of expert knowledge. Often, replicates of sensors installed at the same depth display a certain variation in the readings, which are due to the natural variability of the soil and its natural variation of soil water content. Therefore, 'outliers' must be interpreted carefully, because these readings might represent the natural conditions and are not necessarily caused by false measurements or systems breakdown.

4.4. Related methods

4.4.1. Geophysical methods

Within the last few decades, various non-invasive measurement techniques have emerged, which are also suitable for estimating soil water content not only at a point scale but also horizontally or vertically resolved. The most commonly applied methods are ground-penetrating radar (GPR), electromagnetic induction (EMI), and electrical resistivity tomography (ERT). For the GPR method a nice review is provided by Huisman et al. (2003), and for ERT and EMI we refer to Rubin and Hubbard (2005). Because all methods mentioned do require extended expert knowledge and do not rely on 'classical' sensors, these methods will not be discussed here. In addition to that, a number of other non-invasive techniques allow for continuous and contactless measurements of soil water content dynamics at larger scales (e.g., Global Navigation Satellite System reflectometry, ground-based microwave radiometry, gamma-ray monitoring, terrestrial gravimetry, cosmic-ray neutron monitoring). For information on these approaches, we refer to the work of Bogena et al. (2015) and Vereecken et al. (2014). Because the application of cosmic-ray neutron monitoring is rather straightforward and has become quite popular in the last years we present this technique in more detail in the following section.

4.4.2. Cosmic-ray neutron probe

A recent method to measure integral soil water content at the field or small catchment scale is called cosmic-ray neutron probe (CRNP) (Zreda et al. 2008, Zreda et al. 2012). CRNPs count secondary fast neutrons near the soil surface that are created by primary cosmic-ray particles in the atmosphere and in the soil. Hydrogen atoms in the soil, which are mainly present as water, moderate secondary neutrons on the way back to the surface. Therefore, fewer neutrons escape from moist soils, whereas a larger number of neutrons is able to escape dry soil. This results in a negative correlation between near-surface fast neutron counts and soil water content and enables using CRNP to sense soil water content. The horizontal footprint of the CRNP is not clearly defined. On one hand, a radius of about 300 m which is almost independent of soil water content is reported by Desilets and Zreda (2013). On the other hand, a recent paper suggests that the radius is smaller and strongly dependent on soil water content and atmospheric pressure (Köhli et al. 2015). Furthermore, the measurement depth is strongly dependent on soil water content (~75 cm for dry soils and ~12 cm for wet soils). A further critical point that needs to be considered is the fact that the neutron counts and the sensing volume of the cosmic-ray probe depend on the total amount of hydrogen within the sensor footprint and not only on the hydrogen contained in soil water (Zreda et al. 2012). Additional sources of hydrogen are above- and below-ground biomass, humidity of the lower atmosphere, lattice water of the soil minerals, organic matter and water in the litter layer, intercepted water in the canopy, and soil organic matter. In cases where the amount of hydrogen within these compartments varies with time, this variation must be considered in order to determine soil water content dynamics using CRNP measurements (Bogena et al. 2013). Nevertheless, the applicability of the cosmic-ray neutron probe to measure soil water content has been demonstrated for several different environmental settings, including a coastal site in Hawaii (Desilets et al. 2010), a desert site in Arizona (Franz et al. 2013), an agricultural site with sandy soils near Potsdam, Germany (Rivera Villarreyes et al. 2011), and a low-altitude humid forested catchment in Germany (Bogena et al. 2013). A recent study demonstrated the potential of CRNP measurements for the validation of global soil moisture data products (Montzka et al. 2017).

4.5. References

Augustin, A., Yi, J.Z., Clausen, T., Townsley, W.M., 2016. A study of LoRa: Long range & low power networks for the Internet of Things. Sensors 16 (9), doi: 10.3390/s16091466

Bell, K.R., Blanchard, B.J., Schmugge, T.J., Witczak, M.W., 1980. Analysis of surface moisture variations within large-field sites. Water. Resour. Res. 16 (4): 796–810.

Birchak, J.R., Gardner, C.G., Hipp, J.E., Victor, J.M., 1974. High dielectric constant microwave probes for sensing soil moisture. Proc. IEEE 62 (1): 93–98, doi: 10.1109/proc.1974.9388.

Bircher, S., Skou, N., Jensen, K., Walker, J. & Rasmussen, L., 2012. A soil moisture and temperature network for SMOS validation in Western Denmark. Hydrology and Earth System Sciences 16: 1445–1463. doi: 10.5194/hess-16-1445-2012.

Bircher, S., Andreasen, M., Vuollet, J., Vehviläinen, J., Rautiainen, K., Jonard, F., Weihermüller, L., Zakharova, E., Wigneron, J.-P., Kerr, Y.H., 2016. Soil moisture sensor calibration for organic soil surface layers. Geoscientific Instrumentation Methods and Data Systems. 5: 109–125.

Blonquist, J.M., Jones, S.B., Robinson, D.A., 2005. Standardizing characterization of electromagnetic water content sensors: 2. Evaluation of seven sensing systems. Vadose Zone J. 4: 1059–1069. doi: 10.2136/vzj2004.0141

Bogena, H.R., Herbst, M., Huisman, J.A., Rosenbaum, U., Weuthen, A., Vereecken, H., 2010. Potential of wireless sensor networks for measuring soil water content variability. Vadose Zone J., doi: 10.2136/vzj2009.0173.

Bogena, H.R., Huisman, J.A., Oberdörster, C., Vereecken, H., 2007. Evaluation of a low-cost soil water content sensor for wireless network applications. J. Hydrol. 344: 32–42, doi: 10.1016/j.jhydrol.2007.06.03.

Bogena, H.R., Huisman, J.A., Baatz, R., Hendricks Franssen, H.-J., Vereecken, H., 2013. Accuracy of the cosmic-ray soil water content probe in humid forest ecosystems: The worst case scenario. Water Resour. Res. 49 (9): 5778–5791, doi: 10.1002/wrcr.20463.

Bogena, H.R., Huisman, J.A., Hübner, C., Kusche, J., Jonard, F., Vey, S., Güntner, A., Vereecken, H., 2015. Emerging methods for non-invasive sensing of soil moisture dynamics from field to catchment scale: A review. WIREs Water 2 (6): 635–647, doi: 10.1002/wat2.1097.

Bogena, H.R., Huisman, J.A., Schilling, B., Weuthen, A. and Vereecken, H., 2017. Effective calibration of low-cost soil water content sensors. Sensors 17 (1), 208, doi: 10.3390/s17010208.

Desilets, D., Zreda, M., 2013. Footprint diameter for a cosmic-ray soil moisture probe: Theory and Monte Carlo simulations. Water Resources Research 49 (6): 3566–3575.

Desilets, D., Zreda, M., Ferré, T.P.A., 2010. Nature's neutron probe: Land surface hydrology at an elusive scale with cosmic rays. Water Resources Research 46 (11): W11505.

Entin, J.K., Robock, A., Vinnikov, K.Y., Hollinger, S.E., Liu, S., Namkhai, A., 2000. Temporal and spatial scales of observed soil moisture variations in the extratropics. J. Geophys. Res. Atmos. 105 (D9): 11865–11877.

Famiglietti, J.S., Rudnicki, J.W., Rodell, M., 1998. Variability in surface moisture content along a hillslope transect: Rattlesnake Hill, Texas. J. Hydrol. 210(1–4): 259–281.

Franz, T.E., Zreda, M., Ferre, T.P.A., Rosolem, R., 2013. An assessment of the effect of horizontal soil moisture heterogeneity on the area-average measurement of cosmic-ray neutrons. Water Resources Research 49 (10): 6450–6458.

Gaskin G.J., Miller J.D., 1996. Measurement of soil water content using a simplified impedance measuring technique. Journal of Agricultural Engineering Research 63: 153–160.

Grayson, R.B., Western, A.W., Chiew, F.H.S., Bloschl, G., 1997. Preferred states in spatial soil moisture patterns: Local and nonlocal controls. Water Resour. Res. 33 (12): 2897–2908.

Heimovaara, T.J., Bouten, W., 1990. A computer-controlled 36-channel time domain reflectometry system for monitoring soil water contents. Water Resour. Res. 26 (10): 2311–2316.

Hinnel, A.C., Ferre, T.P.A., Warrick, A.W., 2006. The influence of time domain reflectometry rod induced flow disruption on measured water content during steady state unit gradient flow. Water Resources Research 42: doi: 10.1029/2005WR004604

Hübner, C. et al., 2009. Wireless soil moisture sensor networks for environmental monitoring and vineyard irrigation. 8th International Conference on Electromagnetic Wave Interaction with Water and Moist Substances (ISEMA 2009)), Finland, pp. 408–415.

Huisman, J.A., Hubbard, S.S., Redman, J.D., Annan, A.P., 2003. Measuring soil water content with ground penetrating radar: A review. Vadose Zone Journal 12: 476–791.

Kelleners, T.J., Robinson, D.A., Shouse, P.J., Ayars, J.E., Skaggs, T.H., 2005. Frequency dependence of the complex permittivity and its impact on dielectric sensor calibration in soils. Soil Science Society of America Journal 69 (1): 67–76.

Kizito, F., Campbell, C.S., Campbell, G.S., Cobos, D.R., Teare, B.L., Carter, B., Hopmans, J.W., 2008. Frequency, electrical conductivity and temperature analysis of a low-cost capacitance soil moisture sensor. Journal of Hydrology 352 (3–4): 367–378.

Knight, J.H., 1992. Sensitivity of time domain reflectometry measurements to lateral variations in soil water content. Water Resources Research 28 (9): 2345–2352.

Köhli, M., Schrön, M., Zreda, M., Schmidt, U., Dietrich, P., Zacharias, S., 2015. Footprint characteristics revised for field-scale soil moisture monitoring with cosmic-ray neutrons. Water Resour. Res. 51, doi: 10.1002/2015WR017169.

Lee, H., Seo, D.-J., Koren, V. (2011): Assimilation of streamflow and in situ soil moisture data into operational distributed hydrologic models: Effects of uncertainties in the data and initial model soil moisture states. Advances in Water Resources 34: 1597–1615.

Li, B., Rodell, M., 2013. Spatial variability and its scale dependency of observed and modeled soil moisture over different climate regions. Hydrol. Earth Syst. Sci. 17 (3): 1177–1188.

Martini, E., Wollschläger, U., Kögler, S., Behrens, T., Dietrich, P., Reinstorf, F., Schmidt, K., Weiler, M., Werban, U., Zacharias, S., 2015. Spatial and temporal dynamics of hillslope-scale soil moisture patterns: characteristic states and transition mechanisms. Vadose Zone J. 14 (4); 10.2136/vzj2014.10.0150.

McCready, M.S., Dukes, M.D., Miller, G.L., 2009. Water conservation potential of smart irrigation controllers on St. Augustinegrass. Agricultural Water Management 96 (11): 1623–1632.

Montzka, C., Bogena, H.R., Zreda, M., Monerris, A., Morrison, R., Muddu, S., Vereecken, H., 2017. Validation of Spaceborne and Modelled Surface Soil Moisture Products with Cosmic-Ray Neutron Probes. Remote Sensing 9 (2): 103, doi: 10.3390/rs9020103.

Nagare, R.M., Schincariol, A., Quinton, W.L. & Hayashi, M., 2011. Laboratory calibration of time domain reflectometry to determine moisture content in undisturbed peat samples. European Journal of Soil Science 62: 505–515.

Nemali, K.S., Montesano, F., Dove, S.K., van Iersel, M.W., 2007. Calibration and performance of moisture sensors in soilless substrates: ECH2O and Theta Probes. Scientia Horticulturae 112 (2): 227–234.

Nichol, C., Beckie, R., Smith, L., 2002. Evaluation of uncoated and coated time domain reflectometry probes for high electrical conductivity systems. Soil Science Society America Journal 66: 1454–1465.

Nielsen, D.R., Biggar, J.W., Erh, K.T., 1973. Spatial variability of field-measured soil-water properties. Hilgardia 42 (7): 215–259.

Pepin, S., Livingston, N.J., Hook, W.R., 1995. Temperature dependent measurement errors in time domain reflectometry determination of soil water. Soil Sci. Soc. Am. J. 59: 38–43, doi: 10.2136/sssaj1995.03615995005900010006x.

Qu, W., Bogena., H.R., Huisman, J.A., Vanderborght, J., Schuh, M., Priesack, E., Vereecken, H., 2015. Predicting sub-grid variability of soil water content from basic soil information. Geophys. Res. Lett. 42: 789–796, doi: 10.1002/2014GL062496.

Qu, W., Bogena, H.R., Huisman, J.A., Vereecken, H., 2013: Calibration of a novel low-cost soil water content sensor based on a ring oscillator. Vadose Zone J. 12 (2): doi: 10.2136/vzj2012.0139.

Rejiba, F., Cosenza, P., Camerlynck, C., Tabbagh, A., 2005. Three-dimensional transient electromagnetic modeling for investigating the spatial sensitivity of time domain reflectometry measurements. Water Resources Research 41: doi: 10.1029/2004WR003505

Rivera Villarreyes, C.A., Baroni, G., Oswald, S.E., 2011. Integral quantification of seasonal soil moisture changes in farmland by cosmic-ray neutrons. Hydrol. Earth Syst. Sci. 15 (12): 3843–3859.

Robinson, D.A., Jones, S.B., Wraith, J.M., Or, D., Friedman, S.P., 2003. A Review of Advances in Dielectric and Electrical Conductivity Measurement in Soils Using Time Domain Reflectometry. Vadose Zone J. 2 (4): 444–475.

Robinson, D.A., 2004. Measurement of the solid dielectric permittivity of clay minerals and granular samples using a time domain reflectometry immersion method. Vadose Zone J. 3 (2): 705–713, doi: 10.2136/vzj2004.0705.

Robinson, D.A., Kelleners, T.J., Cooper, J.D., Gardner, C.M.K., Wilson, P., Lebron, I., Logsdon, S., 2005. Evaluation of a capacitance probe frequency response model accounting for bulk electrical conductivity: Comparison with TDR and network analyzer measurements. Vadose Zone J. 4 (4): 992–1003.

Robinson, D.A., Campbell, C.S., Hopmans, J.W., Hornbuckle, B.K., Jones, S.B., Knight, R., Ogden, F., Selker, J., Wendroth, O., (2008), Soil moisture measurement for ecological and hydrological watershed-scale observatories. A review. Vadose Zone J. 7: 358–389, doi: 10.2136/vzj2007.0143.

Romshoo, S.A., 2004. Geostatistical analysis of soil moisture measurements and remotely sensed data at different spatial scales. Environmental Geology 45 (3): 339–349.

Rosenbaum, U., Bogena, H.R., Herbst, M., Huisman, J.A., Peterson, T.J., Weuthen, A., Western, A., Vereecken, H., 2012. Seasonal and event dynamics of spatial soil moisture patterns at the small catchment scale. Water Resour. Res. 48 (10): W10544, doi: 10.1029/2011WR011518.

Rosenbaum, U., Huisman, J.A., Weuthen, A., Vereecken, H., Bogena, H.R., 2010. Sensor-to-sensor variability of the ECH2O EC-5, TE, and 5TE sensors in dielectric liquids. Vadose Zone J. 9: 181–186, doi:10.2136/vzj2009.0036.

Rosenbaum, U., Huisman, J.A., Vrba, J., Vereecken, H., Bogena, H.R., 2011. Correction of temperature and electrical conductivity effects on dielectric permittivity measurements with ECH2O sensors. Vadose Zone J. 10: 582–593, doi: 10.2136/vzj2010.0083.

Rötzer, K., Montzka, C., Bogena, H., Wagner, W., Kidd, R., Vereecken, H., 2014. Catchment scale validation of SMOS and ASCAT soil moisture products using hydrological modelling and temporal stability analysis. J. Hydrol. 519: 934–946.

Rubin, Y., Hubbard, S.S., 2005. Hydrogeophysics. Water Science and Technology Library, Vol. 50. Springer Dordrecht, Berlin, Heidelberg, New York.

Seneviratne, S.I. et al., 2010. Investigating soil moisture-climate interactions in a changing climate: A review. Earth Science Reviews 99 (3–4): 125–161.

Topp, G.C., Davis, J.L., Annan, A.P., 1980. Electromagnetic determination of soil water content: measurements in coaxial transmission lines. Water Resour. Res. 16 (3): 574–582.

Valente, A., Morais, R., Tuli, A., Hopmans, J.W., Kluitenberg, G.J., 2006. Multi-functional probe for small-scale simultaneous measurements of soil thermal properties, water content, and electrical conductivity. Sens. Actuators A: Phys. 132: 70–77.

Vaz, C.M.P., Jones, S., Meding, M. & Tuller, M. 2013. Evaluation of standard calibration functions for eight electromagentic soil moisture sensors. Vadose Zone Journal 12 (2), 1–16.

Vereecken, H. et al., 2008. On the value of soil moisture measurements in vadose zone hydrology: A review. Water Resour. Res. 44: 21.

Vereecken, H., Huisman, J.A., Pachepsky, Y., Montzka, C., van der Kruk, J., Bogena, H., Weihermüller, L., Herbst, M., Martinez, G. Vanderborght, J., 2014: On the spatio-temporal dynamics of soil water content at the field scale. J. Hydrol. 516: 76–96, doi: 10.1016/j.jhydrol.2013.11.061.

Weast, R.C. (ed.) (1986). Handbook of physics and chemistry, 67th ed. CRC Press, Boca Raton, Fl.

Weihermüller, L., Huisman, J.A., Hermes, N., Pickel S., Vereecken, H., 2013. A New TDR Multiplexing System for Reliable Electrical Conductivity and Soil Water Content Measurements. Vadose Zone Journal 12: doi: 10.2136/vzj2012.0194.

Western, A.W., Bloschl, G., Grayson, R.B., 2001. Toward capturing hydrologically significant connectivity in spatial patterns. Water Resour. Res. 37 (1): 83–97.

Wraith, J.M., Or, D., 1999. Temperature effects on soil bulk dielectric permittivity measured by time domain reflectometry: Experimental evidence and hypothesis development. Water Resour. Res. 35: 361–369. doi: 10.1029/1998WR900006.

Yu, X.Q., Wu, P.T., Han, W.T., Zhang, Z.L., 2013. A survey on wireless sensor network infrastructure for agriculture. Computer Standards & Interfaces 35 (1): 59–64.

Zhao, L., Yang, K., Qin, J., Chen, Y., Tang, W., Montzka, C., Wu, H., Lin, C., Han, M., Vereecken, H., 2013. Spatiotemporal analysis of soil moisture observations within a Tibetan mesoscale area and its implication to regional soil moisture measurements. J. Hydrol. 482: 92–104.

Zhu, Q., Liao, K., Xu, Y., Yang, G., Wu, S., Zhou, S., 2013. Monitoring and prediction of soil moisture spatial-temporal variations from a hydropedological perspective: a review. Soil Research 50 (8): 625–637.

Zreda, M., Desilets, D., Ferré, T.P.A., Scott, R.L., 2008. Measuring soil moisture content non-invasively at intermediate spatial scale using cosmic-ray neutrons. Geophysical Research Letters 35 (21): L21402.

Zreda, M., Shuttleworth, W.J., Zeng, X., Zweck, C., Desilets, D., Franz, T. Rosolem, R., 2012: COSMOS: the COsmic-ray Soil Moisture Observing System. Hydrol. Earth Syst. Sci. 16 (11): 4079–4099.

5. Soil matric potential/Soil water tension

Axel Lamparter

5.1. Introduction
5.1.1. General

Forces act on water in soils. These forces determine the amount of water in a soil and the direction of possible water flow. Jointly with the hydraulic conductivity of the soil, these forces quantify the velocity of possible water flow. They define the soil's field capacity, i.e., the volume of water that can be stored in a given volume of soil, against gravitational forces. These forces also define the amount of water, that cannot be extracted by plants (water content at wilting point), and therefore, the proportion of soil water available to plants (plant available water). Soil water contents resulting from forces originating from the solid soil itself are – like a fingerprint – characteristic for each soil.

Different forces act in different directions on the water in the soil, and complicate a detailed description of the force network. Therefore, the potential energy – or just "potential" – of the water in soil is often considered. The potential of the soil water is defined as the energy difference of the soil water as compared to a reference energy state of water. The reference energy state of water is usually defined as pure water (no solutes) at a reference elevation (e.g., the ground water level).

Knowing the potential of the water in a soil is important for calculations of water fluxes/storage. Differences between the total potential (= sum of component potentials) at any location compared to another location are the driving force for water movement between these two locations. To gain a state of equilibrium potential energy in the soil, the value of the total potential tends to be the same at every location in the soil. This equilibrium state is virtually never observed in real soils, as precipitation, evaporation, and removal of water by plant roots (transpiration) prohibit reaching this state.

The total potential of the soil water may be separated into different potentials, depending on their origin:

Gravitational potential, ψz

Earth's gravity is (of course) affecting water in the soil. The gravitational potential, by definition, is the work required to lift a certain quantity of water above an arbitrary reference level. Expressing the gravitational potential on a weight basis (as described in section 5.1.2), the gravitational potential equals the vertical distance of the water to the reference level. The gravitational potential is by definition positive. It increases from the reference level in the soil (e.g., the ground water surface) to the soil surface.

Matric potential, ψm

This potential is caused by the solid soil matrix. Matric forces act as a consequence of free surface/interfacial energies that tend to be minimized in the system by adsorption of water. Usually, solids in soil have high surface (interfacial) free energies (e.g., Ca-montmorillonite 180 m m^{-2}; quartz >400 mJ m^{-2}; Chassin et al. (1986), and Parks (1984) compared to water (72.7 mJ m^{-2}) (Lange's Handbook of Chemistry 1967). To minimize the system's energy state, high energy-surfaces are covered with water (where available). This leads to the formation of a contact angle at the three phase boundary water-solid-air. The adsorption of water on solid surfaces in pores causes the formation of water menisci. The curvature of such a meniscus depends on contact angle and the pore radius. The smaller the pore radius and the contact angle, respectively, the stronger is the curvature of the meniscus and the higher is the pressure difference between water in the pore and the reference bulk water. This pressure difference causes water to rise in the capillary until the hydrostatic pressure equals the described pressure difference. However, adsorbed organic matter on solid surfaces is able to affect interfacial free energies and thus the contact angle may significantly be altered, generally resulting in decreased matric forces (Bauters et al. 2000). By definition, the matric potential is negative. Its value varies – depending on a soil's pore size distribution, its contact angle, and of course its water content – between zero (full saturation) and about -10^7 hPa (very dry soil). Usually, the matric potential is reported as a "pF-value" (*lat. potentia; engl. force*). The pF value is simply the exponent of the absolute value of the matric potential given as a common logarithm (to the base 10).

Osmotic potential, ψo

Soil water generally contains a certain amount of dissolved matter, e.g., salts. Depending on its concentration, water is attracted by this solution. The osmotic potential is by definition the work that is required to extract a certain quantity of pure water from a (salt) solution across a semipermeable membrane. Soils with high salt concentrations in the soil solution have, compared to soils with a low salt concentration in the soil solution, higher water contents at same matric potential. In soils with soil solutions low in salt content the osmotic potential is usually ignored.

Pressure potential, ψp

Air pressure potential, ψpa

The air pressure potential is by definition the potential energy of the water caused by a difference of the air pressure in a soil to a reference air pressure. An air pressure potential can sometimes be measured after heavy rainfall events, when the air in front of the infiltrating water gets compressed and the air pressure increases compared to the atmospheric pressure. Also quick changes in atmospheric pressures may cause pressure differences between atmospheric and soil air pressure. Usually the location of the reference air pressure (e.g., atmosphere) and the location obtained by air pressure in the soil are connected via air filled pores, cracks, etc., leading to an equilibration of the different pressures over time. Once equilibration is complete, the air pressure potential is zero.

Hydrostatic pressure potential, ψpw

The hydrostatic pressure potential is the pressure exerted on the point of interest by an overburden, saturated water column. Again, using the weight of water (cf. section 5.1.2), the hydrostatic pressure potential converts into a distance, i.e., the height of the overburden water column. In practice, the hydrostatic pressure potential assumes values of >0 when water starts (e.g., after heavy rainfall) to pond on a soil surface.

Overburden potential, ψb

Any load, (e.g., a wheel of an agricultural machine) applied to the soil surface of a non-rigid soil causes compaction of the soil matrix. As soon as the soil matrix is compacted and the (low) hydraulic conductivity of the soil prohibits rapid (in comparison to the occurrence of the load) redistribution of the water in the soil, the water potential rises as the water carries part of the load. As the water is redistributed in the soil, ψb decreases and finally reaches 0. The soil shows the same potential – but at a lower water content – than before the load was placed on the soil.

If the load is now removed from the soil and compaction of the soil is irreversible, the total potential does not change and remains at the value that was reached after redistribution. In this case, the water retention curve of the compacted soil has changed. If the load is removed from the soil and the compaction of the soil is reversible, the total potential shows even lower values compared to the initial state, but regains its original value of the potential after water has redistributed (water content rises) in response to unloading.

Combination of different potentials

In soils a combination of different potentials sum up to the total potential, but not all potentials are of the same importance in a specific soil. Thus, there may be no need to measure every component of the total potential. If, for example, a soil is not subject to additional load (overburden), its overburden potential is zero and must not be considered. In soils with very low salt concentrations, the osmotic potential is very low and can be neglected. Table 5.1 schematically lists the main components of the total potential in/at different soils/conditions.

Table 5.1. Main components of the total potential in/at different soils/conditions. The "X" marks where the component potentials make significant contributions to the total potential (incomplete).

Soil/Condition	ψz	ψm	ψo	ψa	ψp	ψb
Rigid soil matrix (unsaturated), low salt concentration	X	X		X		
Rigid soil matrix (saturated), low salt concentration	X			X	X	
Rigid soil matrix (saturated), high salt concentration in soil solution	X	X	X	X		
Non-rigid (swelling/shrinking) soil matrix (saturated), high salt concentration in soil solution	X		X	X	X	X

Certain combinations of different potentials have been shown to be very useful to work with, and therefore, have been defined separately. Additionally, certain sensors are only able to detect combined potentials. Important and frequently used combined potentials are:

Hydraulic potential – combination of matric and gravitational (and gas) potential. Assuming all other potentials are negligible, the hydraulic potential of a soil in equilibrium is zero at every location. As the gravitational potential is, by definition, positive and – when expressed on a weight basis – simply the distance to the ground water. The matric potential then shows the same absolute value – but with a negative algebraic sign. The hydraulic potential is zero and thus, no flow of water occurs.

Water potential – combination of matric and osmotic (and gas) potentials

Tensiometer (pressure) potential – combination of (matric potential, overburden potential, osmotic potential, pressure potential (depending on tensiometer); what a tensiometer measures.

5.1.2. Definitions (SI-units, measurement techniques)

The total potential, or its components, may be expressed on a volume, mass, or weight basis. Resulting dimensions (units) are $M\,LT^{-2}$ [N cm^{-2}], $L^2\,T^{-2}$ [J g^{-2}], or simply a length, L [cm], respectively. In soil science, the potential is often expressed on a weight basis so that the resulting length directly reflects the distance to the ground water (in the equilibrium state).

Measurement techniques to measure the potential in soils generally measure only certain components of the total potential or combinations of these components.

5.1.3. Principles of measuring the potential of the soil water

Generally, the measuring the soil water potential may be achieved by using 3 principles:

1) By measuring the negative (tensiometer) or positive (piezometer) pressure of soil water relative to free water.

2) By measuring water vapor saturation in the gas phase of the soil at the point of interest (e.g., dew point technique), as the relative humidity of the air in the soil is (in equilibrium) dependent on the soil's water potential.

3) After equilibration with a porous reference medium, the matric potential is measured indirectly. Different technical approaches are available based on different parameters of the sensor's porous part, e.g., its electrical permittivity, its water content, its heat capacity, or heat conductivity, etc.

These principles are applied by different methods to determine different components or combination of components of the total soil water potential and are introduced in section 5.2.

5.2. Selection of the appropriate method
5.2.1. Tensiometers

Generally, tensiometers measure the soil's matric potential. The air pressure in the soil is also considered in the measurement if it differs between the point of measurement and the reference pressure (usually ambient air pressure). Additionally, other component potentials such as the hydrostatic/overburden potential are measured by a tensiometer when a saturated water column or some other weight exerts pressure on the soil water or the (non-rigid) soil above the tensiometer, respectively. In most soils of humid regions (generally low salt concentrations), the combination of the matric potential and the gravitational potential equals the total potential and thus determines the water content of the soil and, if the total potential differs between two spatial locations in the soil, the direction of water flow.

Different types of tensiometers are available, with increasing grade of technical features: (i) pipe tensiometers with a water-filled tubular shaft optionally combined with a digital handheld, or combined with fixed vacuummeters (see section 5.2.1.1), and (ii) tensiometers with electronic sensors for permanent monitoring purposes (see section 5.2.1.2ff). The grade of the latter begins with tensiometers where the electronic sensor is mounted on the top of a water-filled pipe (instead of a vacuumeter, see above). The next step are tensiometers with a pressure sensor directly fixed to the porous ceramic cup, and the upgrade of technical features continues with additional temperature measurement, external filling hoses, and ends at high class tensiometers with a self-refilling functionality.

Tensiometers operate comparable to piezometers but are – compared to piezometers – designed to measure under unsaturated soil conditions. To achieve this, tensiometers consist of a water saturated ceramic cup connected to a sealed, rigid water reservoir. Water is able to flow through the pores of the ceramic material, whereas air cannot pass these pores because they are blocked by water. Thus, in- or outflowing water changes the pressure of the water within the tensiometer reservoir. In contact with the soil matrix, the pressure of the water in the tensiometer tends to equalize with the matric potential in the soil matrix. Therefore, water flows through the pores of the ceramic until equilibration between the water pressure within the tensiometer reservoir and the soil is reached (= soil matric potential). Submerged tensiometers are also able to measure a positive pressure. Thus, also, a hydrostatic potential can be measured and the tensiometer used like a piezometer.

The air-entry-value of the ceramic determines the maximum negative pressure that can be adjusted to the reservoir, i.e., when the largest pore in the ceramic empties of water, fills with air and equilibration between the reservoir and the atmosphere takes place. However, usually before reaching the air-entry-value of the ceramic, water in the reservoir is at such a low pressure that it starts boiling at ambient temperature. Consequently, a further reduction of the water pressure in the tensiometer is inhibited by the formation of water vapor – due to the boiling of water – that keeps the water pressure at a constant value. The measurement range of a tensiometer is from about 1000 hPa (depending on pressure sensor) to a negative value that is dependent on the air entry value of the ceramic or the boiling point of the water. The latter is usually restricting the measurement range and is reached at about -800 hPa (at ambient temperatures). The measurement range can be expanded using very pure, degassed water and special tensiometer shafts with polished (very smooth) inner surfaces. Such a system lacks of nucleation sites where water vapor bubbles can form. Thus, water in the tensiometer stays in its liquid form (boiling retardation) expanding the measurement range of the tensiometer. However, due to the contact to the soil water/soil air it cannot be prevented that air and solutes dissolve in the water of the tensiometer over time. As the tensiometer is usually installed in a natural environment, it is subjected to different ambient temperatures. At temperatures below zero degrees centigrade, the formation of ice can destroy the tensiometer. Alternatively, an antifreeze fluid that is not miscible with water (e.g., Decalin; Strebel 1970) can be used for the part of the tensiometer that is exposed to temperatures below zero (near the soil surface). Note, that an antifreeze fluid with a specific weight different from water will theoretically influence the reading.

The accuracy of the tensiometer is dependent on the tensiometer type, and moreover, on the sensor that is used to measure the pressure. U-bend manometers can read precisely in the range ± 1 hPa, manometers and pressure transducers are in the range ± 5 hPa. These values might differ depending on the manufacturer of the device.

5.2.1.1. Pipe tensiometers

Low-cost pipe tensiometers can easily be self-constructed, but they are also commercially available and normally the cheapest tensiometer grade. Such a low-cost tensiometer usually consists of a (transparent) plastic pipe attached to a ceramic cup on the bottom and a silicone plug or septum on the top (Fig. 5.1).

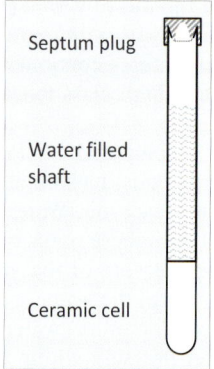 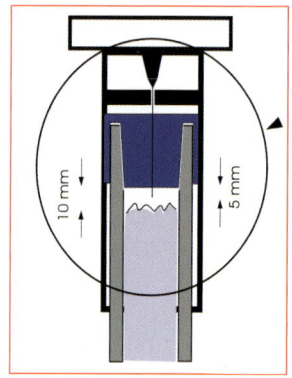

Fig. 5.1. Sketch of a simple pipe tensiometer with water-filled tubular shaft (image source: SDEC, France).

The upper end is connected to a device for the determination of the pressure, i.e. an already fixed vacuum meter or a mobile hand held with a pressure sensor (see section 5.2.1.2).

5.2.1.2. Manual pressure readings

Sealing the tensiometer pipe with an airtight septum, the pressure within the tensiometer can be measured multiple times. With a thin, hollow needle the septum can severally be penetrated and the actual pressure in the tensiometer determined with an attached pressure gauge (low cost).

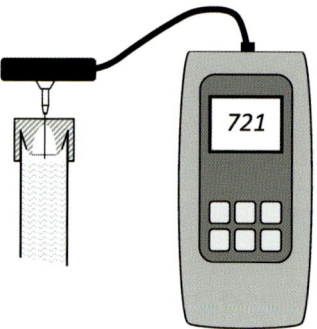 

Fig. 5.2. Sketch of a simple water-filled pipe tensiometer with silicone bung (left) and different reading-out equipment (right) (image sources: SDEC, France).

This kind of manual pressure readings are often time consuming and awkward. However, if a U-bend manometer is used, readings of the pressure can be very precise ($\leq\pm1$ hPa). Evaporation of water from the side of the U-bend manometer exposed to the atmosphere can lead to a lack of soil water due to evaporation over longer time spans. Thus, this way of measuring the matric potential is suited only for a short period of time and pressures close to saturation (about > -100 hPa) as the sides of the U-bend-manometer become inefficiently long.

5.2.1.3. Automated measurement and data recording

5.2.1.3.1. Pressure Sensors

Sensors may be classified according to the type of pressure they measure. Absolute pressure sensors measure the pressure against a vacuum. Atmospheric (or gauge) pressure sensors measure the pressure against an artificial, constant pressure of 1013 hPa (sealed gauge). Most tensiometers measure a differential pressure. Here, the pressure is measured against the current ambient pressure that may vary over time. Hence, quick changes in ambient air pressure might not be accounted for in the pressure reading, and thus, possible changes in the air pressure potential might not be considered. There are different technical approaches to measure the pressure in a system (e.g., capacitive, piezoelectric, optical, thermal, resonant). Tensiometer pressure sensors often use the piezoresistive effect.

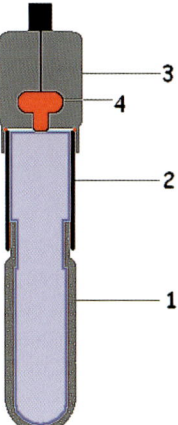

Fig. 5.3. Schematic sketch of a tensiometer with pressure sensor; (1) porous ceramic cell, (2) water-filled transparent pipe, (3) electronics, (4) pressure sensor (image source: wikipedia).

A membrane with embedded electrical resistors is bent depending on the pressure exerted on it. As a consequence, its electrical resistance changes. Using a calibration function, the resistance can be used to calculate the actual pressure. To be calibrated, the tensiometers have to measure the known reference pressure in a system. By plotting the measured data vs. known pressure in an x-y diagram one usually observes a linear relationship between adjusted pressure (known, x-axis) and output signal of the resistor (y-axis). The intercept with the y-axis (pressure = 0) is called the "offset". This offset may change over time and must be checked during longer periods of measurements, while the slope of the function appears in general quite stable over time. The electrical resistance of the resistors often also depends on temperature. As temperatures in soils may change over the course of a set of measurements, the effect of the temperature on the output signal must be considered. When the resistors on the membrane are arranged as a Wheatstone bridge, changes in the ambient temperature do not affect the pressure measurement. Such pressure sensors are often termed "temperature compensated". Piezoresistive pressure sensors need a certain, constant voltage to measure resistance. This voltage is usually about 10 VDC (0–20 V) at a current of 2 mA. The resulting signal is also a voltage in the range of 1–5 V.

5.2.1.3.2. Self-refilling tensiometers

Self-refilling tensiometers can persist under conditions where no measurements can be performed (drought with matric potential < –800 hPa). When conditions return to the range that allows measurement (see section 5.2.1), such tensiometers refill slowly with soil water via an internal vacuum pump and restart measurement. They may be useful in special applications (see section 5.2.4.2), but require additional supply of power for the internal pump. However, their application is relatively costly.

5.2.2. Matrix water potential sensors

As explained in section 5.1.3, water potential may also be measured using principles completely different from tensiometers. Matrix water potential sensors allow measuring a greater range of water potentials than tensiometers. However, readings from these matrix sensors can be of lower accuracy than good tensiometers in certain regions of the matric potential, i.e., where $\Delta\theta/\Delta\psi$ is low. Additionally, hysteresis of the sensor's porous medium itself may further degrade the quality of measurement. Despite this, matrix water potential sensors are increasingly used in environmental science – especially for continuous monitoring applications. This is due to their low maintenance requirements, the very reliable measurement over longer periods of time and because of their ability to cover the entire range of matric potentials with one sensor.

Different sensors with different techniques to determine the water potential are available as listed below.

5.2.2.1. Gypsum blocks

Gypsum blocks are designed to measure the water potential in (periodically) dry soils with water potentials <–800 hPa. Gypsum blocks are small blocks (ca. 2–5 cm^3) consisting of a gypsum matrix with two metal rods incorporated. Small pores in the gypsum matrix equilibrate with the matric potential of the surrounding environment (soil) by in- or outflow of water. Depending on the hydraulic conductivity of the pores and the size of the block, equilibration times can vary. Depending on the water (or better: the gypsum-solution) content in the gypsum block, the electrical resistance between the rods changes. Calibration is required to relate the measured electrical resistance to matric potential in the soil. Gypsum blocks are easily self-made, and are therefore,

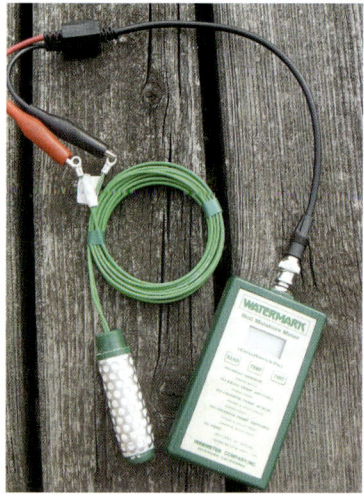

Fig. 5.4. Gypsum block (left) and Watermark-Sensor (right) with handhelds (image sources: Dietmar Rupp, www.lvwo-weinsberg.de)

the most inexpensive method for measuring soil water potentials in periodically dry soils. However, one has to consider that their accuracy and reaction time is low, and that they have a relatively short lifetime due to dilution losses (2–5 years with moreover changing properties). Hence, enlarging gypsum blocks may be useful because this extends their lifetime; but enlargement of the blocks causes the response (to changes in the matrix potential) to become even slower.

An advancement of the gypsum block is the granular matrix sensor (e.g., the Watermark Sensor). Similar to the gypsum block, its electrical resistivity changes with the water content in the granular matrix that is equilibrated with the surrounding soil. A stainless steel mantle physically protects a hydrophilic fabric that covers the granular matrix. The granular matrix sensor has a longer lifetime than gypsum blocks, but are more expensive compared to gypsum blocks. However, they are maintenance-free and have a wide measurement range (from 0 – ~-2400 hPa or even lower). Sometimes, a gypsum block is integrated in the granular matrix sensor for buffering the soil solution, so that the electrical conductivity of the soil solution (or changes in it) does not influence the measurement.

5.2.2.2. Equitensiometer

Equitensiometers consist of a porous material with a known water content/water potential-relation (θ/ψ) (= water retention curve). To measure the matric potential, the porous part of the sensor is equilibrated with the soil's matric potential simply by bringing it in contact with the soil. After equilibration, the water content of the sensor's porous material is measured with the FDR-technique (frequency-domain-reflectometry). FDR makes use of an electromagnetic wave travelling through a wave guide inserted into the sensor's porous medium and being reflected at the end of the wave guide (see section 4.2.4). The difference in the frequency between the injected wave and the reflected wave depends on the dielectric of the surrounding medium (porous equilibrium body) and thus, mainly on its water content. The unknown potential can then be calculated from the measured water content using the known θ/ψ relation.

This technique is used since the late 1990s, and its accuracy depends on how quickly equilibrium between the probe and its surrounding soil is reached (much slower than tensiometers), and on the accuracy of the assumed water retention curve. Accuracy is usually best where $\Delta\theta/\Delta\psi$ is high (the slope of the water retention curve is steep). Manufacturers report measurement ranges of equitensiometers from zero up to –1000 kPa.

Fig. 5.5. Two types of Equitensiometers (image sources: Delta-T devices, Ecomatik).

5.2.2.3. Heat capacity sensors

Heat capacity sensors (also called heat dissipation sensors) use variations in heat capacity and thermal conductivity of porous media, which are controlled by varying amounts of water in soils. Pores in the sensor's matrix equilibrate with the matric potential of the soil. The thermal conductivity (or the specific thermal capacity) of the soil adds up from the static value of the soil matrix itself and the water content-dependent value of the pore water in the matrix.

Heat capacity sensors emit heat pulses into the porous sensor part and measure the temperature changes, which result from this energy input. Some sensors measure the increase of temperature, others measure the decrease (heat dissipation during the cooling phase). Depending on sensor type and manufacturer, some come already calibrated to a number of reference potentials, and thus, are ready-to-use. Others must be calibrated by the operator prior to use. Depending on the measuring principle, the range and whether they are calibrated or not, they widely differ in price, but are more expensive than pipe tensiometers or gypsum blocks.

Heat capacity sensors are designed to work in (periodically) dry soils and/or at sites with a risk of tensiometer damage by frost. This type of sensor usually needs no maintenance and is therefore used for long term monitoring purposes. Its accuracy depends on the measuring method employed, the time required until equilibrium is reached, as well as on the accuracy of the calibration curve.

Examples for heat dissipation matric potential sensors are the pF-Meter, the TensioMark® (measurement range: 0 to –10 000 000 hPa), or the Campbell 229 sensor (measurement range: –100 to –25000 hPa; resolution: ~10 hPa at matric potentials < –1000 hPa)

Fig. 5.6. Types of heat capacity sensors. 1: Tensiomark, 2: pF-Meter, 3: Campbell 229 (image sources: ecoTech GmbH (1, 2), Campbell Sc. (3)).

5.2.3. Automated data recording and storage

If only the current state of the system is of interest (e.g., for irrigation purposes) a single measurement of the potential at a certain location might be sufficient. For applications which require observing the change of the potential over time, manual readings can be very awkward and time consuming. Data loggers are used to automatically measure the pressure at predefined times and store its value. These devices provide robust storage for data and are designed to work under extreme climatic conditions (frost, heat, rainstorms, …). Current generations of data-loggers are remotely accessible by a computer via a mobile communication network. Some are even able to store the measured data in a database, which is accessible over the internet. Disadvantage of an automated measurement of the potential in the soil is, that every sensor requires a power supply (cable) and also for the operation of the data logger electric power is needed. Power may be supplied by batteries. The power supply of batteries is finite, but can be extended by the use of photovoltaic panels – or fuel cells. New generations of sensors are equipped with batteries, and measured data can be transferred wirelessly (e.g., via Bluetooth) to a remote station. Such a setup is costly but pays off quickly in areas where maintenance is very labor intensive.

5.2.4. Decision tree to select the right sensor with respect to the research objective

5.2.4.1. Required number of sensors

Theoretically, the matric potential is a continuous parameter in soils. In equilibrium, the potential – in comparison to the soil water content – does not vary in space, independently of the soil's heterogeneity (e.g., texture, bulk density). However, practically, soils are never in equilibrium, as precipitation, evapotranspiration and other processes inhibit the establishment of stable equilibria. The variability of the potential in soil increases as the heterogeneity of a soil increases, especially as wetting (precipitation) takes place, or soils are drying out. This is due to spatial variation of the hydraulic conductivity, and this effectively leads to a different velocity of adjustment of the potential (equilibration).

Thus, the number of observation points and the distance between them mainly depend on the variability of the boundary conditions, the heterogeneity of the soil, and the absolute value of the potential (dry soils), respectively. Strictly speaking, these measurements cannot be considered as repetitions as the sensors are installed at different locations (which have particular conditions). Hence, to characterize a certain zone (of similar properties) in the soil (horizon), a single measurement is insufficient. Generally, the more measurements of the potential in a soil or horizon, the more detailed is the information gained.

As every measurement is labor intensive and costly, a reasonable compromise is to install 3 sensors per area of interest (e.g., a soil horizon). To obtain information about water flow, the hydraulic gradient must be known, thus at least at two spatial positions (depths) the potential must be known. Continuous measurement of the potential at different spatial locations is necessary where the soil water balance is of interest or if the scientific question requires it. The (scientific) focus of the work finally determines the spatial arrangement, the time resolution of the measurements, and the number of installed sensors required.

5.2.4.2. Expected range of soil water potential

As explained in section 5.2.1–5.2.2, different measurement techniques have different optimal measurement ranges. Either the measurement of the potential in a certain range is simply not possible, or the accuracy of the measurement might not be acceptable for the purpose of the investigation.

In the range from >0 to –800 hPa, tensiometers measure the matric potential most accurately. The accuracy depends (i) on the type of tensiometer used and (ii) on the care which is taken for maintenance. If a particular tensiometer is required because of the expected range of readings, the next criterion for the sensor choice is cost. Simple pipe tensiometers with a silicone plug (see section 5.2.1.1) are the least expensive version with cost of far less than 100 € per sensor. For some applications, e.g., if many sensors are needed to get an impression of the spatial distribution of matric potential at a large site, or for projects with a low budget, pipe tensiometers may be used. Nevertheless, pipe tensiometers have some restrictions: (i) the hanging water column inside (in cm) reduces the instrument's measuring range (see section 5.3.3). (ii) their shaft is partly exposed to the sun, which causes water and air to expand and thus increase the risk of obtaining false readings, and (iii) they are not frost resistant, so their use is limited in winter months.

Tensiometers with electronic sensors in most cases overcome these disadvantages of pipe tensiometers and may in principle be used for continuous monitoring. Units with the sensor directly adjacent to the water-filled ceramics (i.e., a water-filled pipe to the surface is missing) may be used across the year as long as the measuring depth is below the frost zone. They are available in different grades (with corresponding effects on the costs). High-end, self-refilling tensiometers are more expensive than all other mentioned sensors introduced in this chapter. Their measurement range and accuracy is the same compared to conventional tensiometers. They may, however, be useful for sites with periodic drought conditions, where it is of special interest to measure the matric potential with high accuracy in the range >-800 hPa, but information about the matric potential during the dry season (<-800 hPa) is not provided. Another disadvantage of self-refilling units is that they require an additional power supply for the internal pump, which is sometimes difficult to realize in the field.

If the soil is periodically at a matric potential <-800 hPa, and these times are important for the investigation, sensors described in section 5.2.2 are suitable for the measurement of the matric potential. These sensors are also suitable for soils constantly at a matric potential >-800 hPa. As these sensors are also available for different measuring ranges, the expected values must be considered in the selection of the suitable sensor.

Gypsum blocks and Granular Matrix sensors are low-budget sensors and may therefore be deployed in high numbers at moderate costs. They normally measure to a matric potential of -2000 hPa. However, they feature only poor accuracy, especially in the wet region down to -100 hPa. Where matric potentials <-2000 hPa are expected, or where these low-budget sensors are not suitable due to their short lifetime or poor accuracy, heat capacity sensors may be the right choice. Following the specifications given by the manufacturers, they feature large differences in terms of their measuring range, life-time, geometry, accuracy, and price.

5.2.4.3. On-site infrastructure and other measurements

Depending on the infrastructure available at the site of measurement, different measurement techniques are reasonable to conduct. Availability of electrical power, available ports at a data logger or simply an already excavated soil profile, which simplifies sensor installation, may influence the decision what measurement technique to use, the type of sensor, or the number of repetitions of each measurement.

5.2.4.4. Soil type and soil horizons, soil texture, obstacles (roots, skeleton)

Unfavorable is the installation of sensors in soils/horizons with high stone contents. Stones can restrict or even preclude the installation of boreholes or cause bad contact between sensor and soil matrix. Sensor installation in soils high in clay content can also be problematic, as these soils may swell and shrink as their water contents changes. Under such conditions, sensor-shafts may be destroyed (by bending or breaking) and/or sensors may lose contact to the soil matrix. Indurated soil horizons (e.g., petroplinthic horizons) may complicate the installation of sensors. In these cemented horizons, a good contact between sensor and matrix can only be achieved by the use of a silt/water suspension, which is poured into the borehole while the sensor is installed.

During dry seasons, plants (roots) have been observed to use the water reservoir in the tensiometer as water supply. As a result, a tensiometer reading may (i) rather reflect the suction of the roots than of the soil, and/or (ii) plant root tomentum around the tensiometer cup may increase the response time of the tensiometer. This problem increases with root density in the soil and is generally limited to measurements in the main root zone. Depending on the risk that plant roots affect the sensor readings of tensiometers, using matrix sensors (section 5.2.2, no water source!) should be considered.

5.3. Implementation

5.3.1. Sensor installation procedure

5.3.1.1. Installation of tensiometers

Before installing a tensiometer, all porous parts of its sensor must be submerged in water for some days (2–3 days). This procedure produces completely wettable surfaces and ensures that the porous part of the tensiometer is air-free (= completely water saturated). Thus, no resistance to water flow due to air pockets is to be expected and equilibration time of the tensiometer is optimal. Additionally, the air-entry value is – due to the completely wettable surfaces – at its maximum. The air-entry value is the pressure at which the largest water filled pores are emptied so that air may enter the tensiometer. This air-entry quickly equilibrates the pressure in the tensiometer with the atmospheric pressure, terminating the measurement.

Tensiometers are generally installed in boreholes in the soil. The radius of the bottommost centimeters of the borehole should be slightly smaller than radius of the ceramic cup of the tensiometer to assure a tight fit and thus intimate contact with the surrounding soil. This can be achieved by using a special auger for sensor installation. Alternatively, a silt/water suspension may be used, which is filled into the borehole after installation of the ceramic to achieve good contact with the soil matrix. This procedure is recommended in soils with a high amount of cobbles. Before installation, the saturated porous material of the sensor should *not* be exposed to the atmosphere, as due to evaporation, the air-entry value of the porous material can quickly be reached. In this case, the tensiometer needs to be filled with water again to remove air bubbles before installation into the soil. Therefore, ready-to-use sensors should always be stored with a wet tissue wrapped around them or immersed in water.

Tensiometers should be installed into the soil at an angle (to the vertical direction) between >0° and <90° (best at 45°, when installed from soil surface). This reduces the risk of (rain-) water preferentially flowing down the shaft and leading to incorrect measurements. Additionally, after installation of a sensor, a rubber ring around the tensiometer shaft should cover the borehole in order to prevent surface water from entering the borehole.

5.3.1.2. Installation of other sensors

Sensors as were described in section 5.2.2 are usually installed in the soil similar to tensiometers, i.e., either from the soil surface or from a pit or trench, preferably at an angle (e.g., 45°) to the surface (see section 5.3.1.1). The different shapes of most sensors (other than tensiometers) make it difficult or impossible to install them without resorting to a slurry of water and silt or soil. This is time consuming, but often allows verifying a good contact between sensor and soil. Additionally, the cable connecting the sensor with the data logger/power supply can be positioned in the soil, in a way that prevents fast draining water to be preferentially led to the sensor and compromising the measurement.

5.3.2. Measurement and sensor maintenance

Tensiometers must be filled with degassed water. This applies to both pipe tensiometers with water-filled shafts and for electronic tensiometers, which have water only in the ceramic cell. Degassing the water prevents the formation of air bubbles in the tensiometer that might dampen the tensiometer signal. Water may be degassed by boiling it and/or by exposing the water to a low air pressure (< -950 hPa) for some time. Air bubbles, which may form after some time of usage, must be removed periodically. Pipes (as well as refilling hoses) must be protected from radiation (e.g., the sun) as warm water tends to form gas bubbles. Additionally, solar radiation might decrease the lifetime of the tensiometer shaft when it is made of plastic.

Measurement of the matric potential should only be conducted when the sensor is in equilibrium with the surrounding soil matrix. After installation, equilibration will take some time (at least hours, or even days). Depending on the sensor type and/or the installation method, this may have different reasons. When a water-filled tensiometer is installed in a relatively dry soil without adding a slurry during the installation process for better contact (and thus with no additional water) between sensor and soil, the tensiometer will emit some of its internal water to the dry soil during equilibration. Hence, the sensor moistens the soil surrounding it, and thus, affects the parameter to be measured. It is therefore necessary to wait for the first reliable reading until the soil has reached its initial matric potential again. If a slurry for good contact between sensor and soil is used, water from the slurry also alters the potential of the soil. Hence, it is important not to use the readings conducted too short after installation. By observing the first readings over a period of hours/days gives an impression of when the readings will approach the initial, the "real" value to be measured.

Roots favor the presence of tensiometer cups, as they are a reliant water supply during dry seasons (although quickly exhausted). Where tensiometers are installed in soil horizons of high root density, care should be taken that roots will not totally cover the ceramic cup, and so extend the time of the instrument to respond to changing matric potentials in the soil (see also section 5.2.4.4).

Contamination of the sensor tip with organic matter should be avoided as organic remnants cannot only block pores but, depending on the type of organic substance, might turn the surface of the porous material hydrophobic after drying and storage. If a contamination has occurred, cleaning the sensors' porous material before installation is highly recommended (follow the manufacturers' instructions). Cleaning might change the known θ/ψ-relation of the porous medium (when sensors such as those described in section 5.2.2 are used), which is important for the determination of ψ. It might also alter the air-entry value (mentioned above) of the porous material of a tensiometer. Therefore, porous parts of the sensor should only be handled with gloves to prevent contaminating them with organic substances (e.g., grease). Porous materials should not be cleaned with a detergent, which contains surface-active substances (surfactants). These substances may decrease – when not thoroughly rinsed off before installation in the soil – the surface tension of the water in the porous material, thus falsifying the measurement. Hydrogen peroxide may be used (if recommended by the manufacturer) to eliminate organic substances in/on the ceramic cup. Inorganic contaminants may be eliminated in an ultrasonic bath (if recommended by the manufacturer).

5.3.3. Raw data processing and measurement error handling

As the water-filled shaft of the pipe tensiometers (section 5.2.1.1) acts like a hanging water column, the angle of installation influences the measured value (see Eq. 5.1). If the angle of installation or the hanging water column, respectively, is not considered in the calibration function (electronic tensiometers are calibrated in this position), the offset of the measurement must to be corrected for as follows:

$$WC = \sin(\alpha)*T \qquad (5.1)$$

Where WC is the length of water column (L), T is the length of water filled tensiometer shaft (L), α is the angle of installation of the shaft relative to the horizontal (deg.).

The measured values from the tensiometer are the sum of WC and the matric potential. The matric potential is in units of pressure (usually in hPa) – therefore the value of WC (in units of cm water column) must be converted from the dimension of a length to the same unit of pressure. This conversion depends on the density of the water (temperature) and on the acceleration (g) due to gravity (geographic position). The following approximation holds with an adequate accuracy: value in cm water column * 0.9807 = value in hPa).

If the pressure transducer is measuring against atmospheric pressure, aeration of the pressure transducer must be provided. Usually, this is realized by an extra air-filled hose, that connects the pressure transducer to the atmosphere. This hose must be kept clean and dry at all times to ensure a pressure equilibration of the transducer with the atmosphere.

Equation 5.1 may also be used to determine the depth of the borehole if the tensiometer is installed at an angle >0° and <90° to the horizontal direction. In this case WC denotes the installation depth, T is the length of the buried shaft of the tensiometer (= length of the borehole), and α is the angle of the borehole to the horizontal direction.

Pressure transducers of tensiometers, using piezoresistive technology, are – due to the use of a Wheatstone bridge – temperature compensated. Using pressure transducers, which are not temperature compensated, an influence by varying ambient temperature can sometimes be measured. This is avoided by measuring the signal of a *second* pressure transducer exposed to the same ambient temperature changes but not connected to a tensiometer. Differences in the measured signal over time usually correlate with changing ambient temperature. Therefore, these differences of the two signals may be used to eliminate the temperature effect in the measured data.

Some of the non-tensiometer-sensors display their readings as "pF-value" instead of hPa or kPa. Please note that averaging pF-values leads to other results than averaging Pa-values. Example: mean of the three readings 100 hPa, 200 hPa and 300 hPa is 200 hPa. Averaging the same matric potentials as pF-values (= pF 2,0, pF ≈ 2,3 and pF ≈ 2,48) will result in a mean pF ≈ 2,26 which corresponds to ≈ 182 hPa (instead of 200 hPa). Hence, to compare different parameter readings, hPa-readings need to be logarithmised or pF-readings de-logarithmised prior to comparison.

When using tensiometers, special care should be taken after dry periods, when the tensiometer has dried up and has stopped measuring. First, note that refilling empty tensiometers only makes sense when the matric potential of the soil has become greater than –800 hPa by precipitation or irrigation, because otherwise the tensiometer will dry out again immediately. Secondly, identify those periods with unreliable readings to tell them apart sharply from the other (= good) data. This may be complicated and time-consuming in long-term experiments, but greatly influences the data and ultimately their interpretation. Especially, where such data are used in models, data gaps and rapid changes of readings can lead to errors and unstable models.

Measuring the soil water potential in swelling/shrinking soils can be difficult. Due to shrinkage, the sensor may lose contact with the soil matrix. Sensors with a shaft may even be destroyed due to swelling/shrinking. In such soils, using sensors without a shaft is recommended.

5.4. Related Methods

5.4.1. Field: Suction cups and plates for soil water sampling

Suction cups are used to extract soil water samples (see section 6.2). Chemical properties or solute concentrations may be determined in the extracted water. A slightly higher suction (lower negative pressure), than the matric potential in the soil, is applied to a water saturated filter material connected to a reservoir. This small difference in the potential forces soil water to flow into the reservoir. This principle works as long as the suction within the suction cup or plate is lower than the air entry value of the filter (otherwise, air is sucked into the reservoir). Depending on the scientific focus of the experiment, the strength of applied suction may be controlled by a tensiometer installed in the vicinity of the soil water extraction device. The applied suction is adjusted to a value slightly higher (more negative!) than the values measured by the tensiometer. This ensures that only water is sampled that would otherwise have been transported into deeper regions of the soil.

5.4.2. Laboratory: Determination of the total potential by water vapor sorption/dewpoint potentiometry

Water vapor sorption/dewpoint potentiometry is a quick and easy laboratory method to measure the soil's total potential. The relative humidity in the air phase adjacent to a soil sample depends on the total potential of the soil and the ambient temperature. Controlled changes in temperature may be used to determine the dew point (rel. humidity is 100%). With this information, one is able to calculate the total potential of a soil. Additionally, the soil's water content can be determined gravimetrically (after drying the soil in an oven at 105 °C). This quick and simple method yields information about the θ/ψ-relation at the very dry end of the water retention curve.

5.5. References

Bauters, T.W.J., Steenhuis, T.S., DiCarlo, D.A., Nieber, J.L., Dekker, L.W., Ritsema, C.J., Parlange, J.-Y., Haverkamp, R., 2000. Physics of water repellent soils. Journal of Hydrology 231–232: 233–243.
Chassin, P., Jounay, C., Quiquampoix, H., 1986. Measurement of the surface free energy of Calcium-Montmorillonite. Clay minerals 21: 899–907.
Lange's Handbook of Chemistry, 1967. 10th ed., pp. 1661–1665.
Parks, G.A., 1984. Surface and interfacial free energies of quartz. Journal of Geophysical Research 89 (B6): 3997–4008, doi: 10.1029/JB089iB06p03997
Strebel, O., 1970. Messung der Bodenwasserspannung mit Hg-Manometer-Tensiometern bei Lufttemperaturen unter 0 °C. Zeitschrift für Pflanzenernährung und Bodenkunde 126: 111.

6. In-situ soil water sampling

Lutz Weihermüller

6.1. Introduction

The knowledge of the quality and quantity of soil pore water in the unsaturated zone is essential for a wide range of practical and scientific questions, including ecology, water management, agriculture, forestry and environmental protection, pesticide registration, fate of xenobiotics, and monitoring of disposal from mining and industries. The rapid development of analytical methods during the past decades offers the possibility of a comprehensive qualitative and quantitative documentation of the solutes dissolved in the soil water. This requires an optimal sampling system to avoid bias and artifacts from contamination and changes of the extracted soil water. Furthermore, new developments in data analysis, modeling, and monitoring of water and solute flow by e.g. non-invasive methods (see chapter 4) increased the requirements of sampling and changed needs of the experimental design.

Suction cups, porous plates, pan lysimeters, capillary wicks, and lysimeters are widely used for in-situ monitoring of the soil solution. The advantage of the aforementioned methods is the high spatial and/or temporal resolution of monitoring the movement of the dissolved substances in the water phase. In contrast to the liquid extraction of sampled soil cores, concentration variations over time can be observed at exactly the same location in the soil profile, and therefore, the impact of spatial heterogeneity is less important compared to the extraction of pore water from soil cores. Moreover, large experience has been gained regarding potential sources of errors of in-situ soil sampling.

Irrespectively of the choice of the sampler or the material the sampler is build up, all extraction devices provide two different information. First, the extracted amount of water (liter or mm) can provide information about water leaching in the soil, and second the concentration of substances in the extracted water (g l^{-1}) provide information about the concentration or the mass flow through a defined plane within the soil system.

Over the last decades, a wide range of new materials were developed or adapted for the different sampling techniques for all kind of scientific applications, which makes it difficult to discuss all sampler materials and samplers in this chapter. Therefore, we will introduce the different samplers with advantages and disadvantages and provide a general overview of the different materials in use at the end of the chapter.

6.2. Suction cups

Various terms are in use to describe the same sampling device, such as porous tube, deep pressure vacuum lysimeter, vacuum extractor, porous candle, porous cup, suction cup, suction lysimeter, or suction probe (Duke and Haise 1973, Grossmann and Udluft 1991, Krone et al. 1951, Parizek and Lane 1970). In the following, the term suction cup for the whole sampling device will be used. The principle of the suction cup was first described by Briggs and McCall (1904). Since then, suction cups were widely used in different studies to collect soil water for scientific purposes, whereby the systems might differ in length, diameter, and material. In general, suction cups consist of a conical or cylindrical porous body connected to one (or two) extraction tube(s) (Fig. 6.1), which will be connected to a sampling bottle. Via the sampling bottle, a suction will be applied to the cup. For the porous body porous ceramics, sintered glass or metals, or membranes are nowadays in use (Weihermüller et al. 2007).

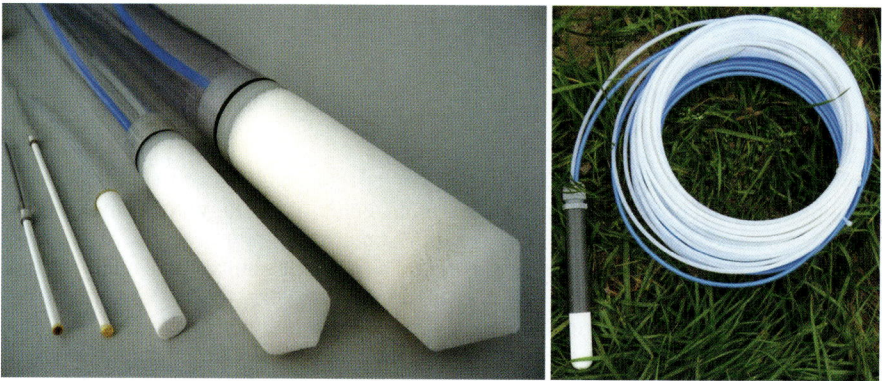

Fig. 6.1. Suction cups of different size and shape (Photos: ecoTech GmbH, Bonn).

6.2.1. Operation of suction cups

For the operation of a suction cup, a negative pressure (often referred as vacuum) has to be imposed to the suction cup by a vacuum pump. The optimal height of applied suction to the cup and the optimal operation mode are still under debate (McGuire and Lowery 1994, Brandi-Dohrn et al. 1996, Weihermüller et al. 2005, Weihermüller et al. 2011). In general, the suction to be applied to the porous cup depends on the soil type, the specific amount of water required for analysis, the actual soil water content, and the time of applied suction (Warrick and Amoozegar-Fard 1977, Weihermüller et al. 2005).

For the soil water extraction with porous cups three different operation modes are reported:

i) For the discontinuous operation mode, water collection is performed during selected short-time intervals at predefined pressure. This operation mode is used to indicate the presence of solutes at specific points of time (Linden 1977). The advantage is the small temporary disturbance of the natural flow field and low maintenance. The disadvantage, on the other hand, is a non-permanent flow through the cup material, which can result in high sorption of the target substance due to longer contact times of the solute and the porous material. Therefore, it is useful to discharge the first water sampled. It is also discussed that the discontinuous operation mode might not be able to record short-time events such as rapidly changing concentrations of solutes caused by heavy rainfall and preferential flow (Jury and Flühler 1992), whereby Weihermüller et al. (2011) showed that both discontinuous and continuous operation mode will provide the same information as long as discontinuous sampling will be performed on relatively short intervals.

ii) The second possibility is the discontinuous operation mode with falling head. Therefore, the sampling will be only periodically, similar to the discontinuous mode, but the pressure will not be kept constant over time. Classically, this operation mode will be fully manual by applying the suction via mobile vacuum pumps to the sampling bottles or via a vacuum container connected to the suction cup. As a consequence, the water will be collected over time while the vacuum is decreasing. A major drawback of this operation mode is that the pressure drop over time will not be exactly known. Additionally, only small amounts of water are often extracted.

iii) The last operation mode is the continuous operation mode, where a constant potential gradient will be applied, which will be calculated based on information about the ambient pressure of the soil measured by reference matric potential sensors (see chapter 5). Additionally, a predefined pressure offset is added to the measured matric potential sensor value. The advantages of the matric potential controlled continuous operation mode are the permanent extraction of soil water, and therefore, a relatively accurate assessment of the drainage pattern (Magid et al. 1992, McGuire and Lowery 1994). Additionally, the small withdrawal of water per time unit will minimize the changes of the natural water flow pattern (Grossmann and Udluft 1991). The continuous water flow might also reduce sorption processes due to shorter contact times of the solute with the porous material and only low potential gradients are necessary to collect adequate amounts of water for chemical analysis. On the other hand, permanent extraction is relatively expensive because reference matric potential sensors and a control unit have to be installed. Additionally, the continuous pressure head gradient might create potential preferential flow paths towards the cup (Weihermüller et al. 2005) and high effort for maintaining the system is required. In fact, most available systems refer to only one reference sensor for a whole arrangement of suction devices, working on an area of several square meters or even greater dimensions. As a result of spatial heterogeneity of matric potentials in the measuring and extraction layer, the single reference matric potential produces an unknown error. So, besides the fact, that those systems are cost- and maintenance-intensive, they give the impression of having a high-sophisticated technical solution for calculating solute freights.

Finally, changes of sample composition might occur due to relatively long storage of the extracted water under field conditions.

6.2.1.1. Continuous data recording or single observations

Because suction cup data generally integrate over a certain time period (depending on operation mode and exchange of sampling bottles), data with high temporal resolution are rare, especially in field conditions, where daily exchange of the sampling bottles is not feasible. Additionally, the total amount of extracted water depends on the operation mode and needs to be sufficient for chemical analysis. As mentioned above, discontinuous and continuous operation modes are feasible, whereby both seem to be appropriate to capture the solute fluxes as long as discontinuous sampling will be performed at relative short time intervals. These intervals depend on the hydraulic conditions of the site, whereby sampling and exchange of sample bottle (extract) should be in smaller time intervals during periods of high leaching (e.g., wet periods) and can be longer during drier conditions. In all cases, overflow of the sampling bottles should be avoided.

6.2.1.2. Range of application

In general, water extraction by suction cups is limited by the vacuum which can be applied to the soil. As the general vacuum limit which can be applied is around 1000 cm (~1 bar) water cannot be extracted if the matric suction of the soil exceeds this threshold. Especially, for coarse soils (e.g., sand) this threshold can be easily reached at dry soil conditions especially close to the soil surface. For finer soils, this threshold will be only reached under extremely dry conditions very close to the soil surface. Additionally, dryer soils have lower unsaturated hydraulic conductivities leading to slower sampling and less water extraction per time unit.

For coarse soils such as sand the suction applied to the cups should be in general lower as for fine soils because large suctions will desaturate the soil in direct vicinity of the cup, and therefore, reduce the amount of water sampled dramatically. This is especially the case for the discontinuous operation mode, where normally high suctions are applied for a relative short time interval.

6.2.2. On-site infrastructure and other measurements

The on-site infrastructure necessary for operating suction cups greatly depends on the suction cup operation mode. The simplest way to operate suction cups is the *discontinuous operation with falling* head because i) mobile vacuum pumps can be used for the extraction, or ii) the suction can be applied via vacuum containers, which can be evacuated at the laboratory. Therefore, no equipment such as permanently installed pumps is needed. Additionally, no further data such as soil matric potential (see chapter 5) are needed for the operation itself. In consequence, this operation can be performed at any field even without power supply.

For the *discontinuous operation mode,* a constant pressure will be applied over a predefined time interval. To do so, at least a vacuum pump is needed which will run either on battery or permanent power. As for the discontinuous operation with falling head, no additional sensors have to be applied for water extraction. The most sophisticated operation mode is the *continuous operation mode*, whereby the suction is controlled by matric potential sensors and a control unit. Therefore, this setting needs permanent power supply, a matric potential sensor setup, and a control unit.

6.2.2.1. Soil type: Depth of interest, soil texture, obstacles: roots, skeleton

In general, suction cups can be installed and operated at all locations and soil types. The installation depth is only limited by the accessibility (e.g., depth of access trench) and the technical limitations of the suction pump (e.g., lift height of the vacuum pump). To ensure a proper function, good hydraulic contact between the soil and the cup should be ensured while installation by preparing a narrow borehole or the use of slurry. This applies in particular for soils rich in skeleton or rocks.

6.2.2.2. Other constrains

A general overview of different experimental designs and constrains is already provided in section 1.5. Because water and solute movement in the soil is normally a rather slow process, and deep percolation (>1 m) might take several months to even years, especially for reactive (sorbing) substances, the experimental phase should be planned to be long enough to capture the full details (e.g., breakthrough curve) of the substance under investigation. To get a rough estimate of the minimal time period required, hydrological and solute transport modelling might help. Additionally, directly after installation a so-called "first flush" is often observed, characterized by an extreme increase of solute concentrations over a short time period. The reason for that can be found in the installation process and the associated disturbance and mechanically mixing of the soil, which changes e.g. the oxygen concentration and might release considerable amounts of organic matter and other soil compounds. Hence, suction cups should be installed several weeks before the start of the experiment to equilibrate with the soil solution. This equilibration time can also reduce sorption artifacts by the cup material and can be used to check proper installation. It is recommended to operate the suction cups even in this adaption/equilibration phase, but the samples should be either rejected or the results should be used carefully for a differentiation of the installation-influenced concentrations (first flush) from the reliable results of the study period.

6.2.3. Application of suction cups

6.2.3.1. Experimental layout

The installation of the suction cups (horizontally, diagonally, vertically) is mainly determined by the boundary conditions of the experiment (e.g., accessibility), whereby horizontal installation from an access trench is generally preferable because it reduces the possibility of vertical preferential flow of water and solutes along the suction cup shaft or within the backfilled material.

In contrast to other sensors, the sampling volume of suction cups is undefined 3-dimensional, and its geometry and space mainly depends on i) the hydraulic conditions in the vicinity of the cup and ii) the duration of sampling. Moreover, the extraction volume will vary under different water saturation levels, so the composition of the collected sample can be dominated by short-term preferential flow events (Wessel-Bothe 2002). Hence, the question of replicates highly depends on the heterogeneity of the field under study, whereby larger heterogeneity will request more replicates. In general, sampling volumes (in terms of sampled soil volume) are small, and therefore, only represent a small volume of the soil. It has been shown by Weihermüller et al. (2011) that the total number of suction cups needed for reliable estimation of solute fluxes in one soil depth might exceed the classical number of replications ($n = 3$) due to soil heterogeneity, whereby the precise number of needed cups remains unknown.

6.2.3.2. Self-construction of suction cups

Theoretically, suction cups can be self-constructed, whereby the most critical part of the sampler are the porous cups, which normally do have a large air entry value, and therefore, require special materials not always available at the free market. Additionally, material should be free of contamination – this should include tubings and glues.

6.2.3.3. Sampler material

The material of suction cups should be selected carefully with respect to the substance under investigation. The tendencies of suction cups to sorb organic and inorganic compounds have been widely discussed in literature by Wagner (1962), Hansen and Harris (1975), McGuire et al. (1992), and Wessel-Bothe et al. (2000), amongst others. As stated before, the development and use of new materials such as sintered metals and glass or membranes allow minimizing sorption effects or the contamination of the soil solution samples by the material itself. An overview of the sorption tendencies with respect to specific substances is provided in Table 6.1. Additionally, suction cups should be cleaned and tested before installation to avoid contamination. The cleaning procedure depends on the materials, and cleaning instructions should be provided by the manufacturer. Testing should include a vacuum test, whereby the suction cups should be settled into a bucket of water and a vacuum should be applied. If the vacuum will not drop over short time even after taking them out of the water, the cups are tight and can be installed. Because most suction cups are made out of sintered materials (glass, metal, or ceramic) the hydraulic conductivity might differ greatly between cups, and therefore, also the extracted amount of water in the field. To determine the saturated hydraulic conductivity of the porous material, the suction cup will be again settled into a bucket of water (keep the water level as constant as possible) and to all cups the same vacuum will be applied. The total amount of sampled water per time units will provide information about the variability in hydraulic conductivity.

Table 6.1.

Test substance	Potential artefacts	Materials with good documented or expected suitability	Materials with limited suitability, documented or expected	Provisions	Literature
aluminium (Al(OH)xn-x)	sorption, precipitation by increase in pH	PA, PE, PTFE	stainless steel oxide ceramic	dual chamber suction cup	Hädrich et al. (1977), Suarez (1986 and 1987), Kaupenjohann and David (1996), Goyne et al. (2000)
ammonium (NH_4^+)	nitrification, N-assimilation	stainless steel, glass†, oxide ceramic, nylon, PE, PTFE, PVC		short sampling intervall	
lead (Pb^{2+})	sorption contamination	PA, PE, PTFE	stainless steel oxide ceramic PVC		Hädrich et al. (1977), Grossmann et al. (1985), Grossmann et al. (1990), Wenzel and Wieshammer (1995), Wenzel et al. (1997), Rais et al. (2006)
cadmium (Cd^{2+})	sorption contamination	PA, PE, PTFE	stainless steel oxide ceramic PVC		Hädrich et al. (1977), Grossmann et al. (1985), Grossmann et al. (1990), Wenzel and Wieshammer (1995), Wenzel et al. (1997), Rais et al. (2006)
iron (Fe^{2+})	precipitation by oxygen contact or increase in pH	glass, Nylon, PE, PTFE, PVC		dual chamber suction cup, adding of acids in sampling bottle	Hädrich et al. (1977), Suarez (1986 and 1987), Schwartz and Mielich (1993)
dissolved organic bound nutrients (e.g., DON, DOP)	specific sorption, contamination from solvents and flexibilizers	stainless steel, glass, Nylon, PE, PTFE, PVC	oxide ceramic adhesives elastomeric bonds	adhesive and elastomeric bond free samplers	
dissolved organic carbon (DOC)	specific sorption, contamination from solvents and flexibilizers	stainless steel, glass, Nylon, PE, PTFE, PVC	odixe ceramic adhesives elastomeric bonds	adhesive and elastomeric bond free samplers	Guggenberger and Zech (1992), Wessel-Bothe et al. (2000), Siemens and Kaupenjohann (2003)
potassium (K^+)	cation exchange	stainless steel, glass, (oxide ceramic), Nylon, PE, PTFE, PVC			Grover and Lamborn (1970), Hädrich et al. (1977)
calcium (Ca^{2+})	cation exchange, precipitation of carbonate	stainless steel, glass, (oxide ceramic), Nylon, PE, PTFE, PVC		dual chamber suction cups	Grover and Lamborn (1970), Hädrich et al. (1977), Schwartz and Mielich (1993)
carbonate (H_2CO_3, HCO_3^-, CO_3^{2-})	degassing of CO_2	stainless steel, glass, oxide ceramic, Nylon, PE, PTFE, PVC	glass fiber wicks	dual chamber suction cups; additional soil gas sampling and corrections	Suarez (1986 and 1987), Grossmann et al. (1988), Kaupenjohann and David (1996), Goyne et al. (2000)
copper (Cu^{2+})	sorption, contamination	PA, PE, PTFE	stainless steel, oxide ceramic, PVC		Hädrich et al. (1977) Grossmann et al. (1985) Grossmann et al. (1990), Wenzel and Wieshammer (1995), Wenzel et al. (1997), Rais et al. (2006)
magnesium (Mg^{2+})	cation exchange, precipitation of carbonate	stainless steel, glass, (oxide ceramic), Nylon, PE, PTFE, PVC		dual chamber suction cups	Grover and Lamborn (1970), Hädrich et al. (1977), Schwartz and Mielich (1993)

Table 6.1 (continued)

Test substance	Potential artefacts	Materials with good documented or expected suitability	Materials with limited suitability, documented or expected	Provisions	Literature
manganese (Mn^{2+})	precipitation by oxygen contact or pH enrichment	glass, Nylon, PE, PTFE, PVC		dual chamber suction cups, adding of acids in sampling bottle	Hädrich et al. (1977), Suarez (1986 and 1987), Schwartz and Mielich (1993)
sodium (Na^+)	cation exchange	stainless steel, glass, (oxide ceramic), Nylon, PE, PTFE, PVC			Grover and Lamborn (1970), Hädrich et al. (1977)
nickel (Ni^{2+})	sorption, contamination	PA, PE, PTFE	stainless steel, oxide ceramic, PVC		Hädrich et al. (1977), Grossmann et al. (1985), Grossmann et al. (1990)
nitrate (NO_3^-)	nitrification, N-assimilation	stainless steel, glass, oxide ceramic, Nylon, PE, PTFE, PVC		short sampling interval	
pesticides, xenobiotics, hormons	sorption, contamination from solvents and flexibilizers, volatilization	glass, stainless steel	oxide ceramic, PE, PVC, (PTFE)	adhesive and elastomeric bond free samplers, dual chamber systems	Wood et al. (1981), Wessel-Bothe et al. (2000)
phosphate ($H_xPO_4^{x-3}$)	specific sorption by ligand exchange	stainless steel, glass	oxide ceramic		Hansen and Harris (1975), Hädrich et al. (1977), Bottcher et al. (1984)
poly-chlorinated biphenyl (PCB)	sorption	glass, stainless steel	oxide ceramic, PE, PVC, (PTFE)	adhesive and elastomeric bond free samplers	
polycyclic aromatic hydrocarbon (PAH)	sorption, volatilization	glass, stainless steel	oxide ceramic, PE, PVC, (PTFE)	adhesive and elastomeric bond free samplers, dual chamber systems	Wood et al. (1981)
protons (H^+, pH-value)	degassing of CO_2	stainless stee, glass, oxide ceramic, Nylon, PE, PTFE, PVC	specific fiber glass wicks	dual chamber sampling systems, additional soil gas sampling and corrections	Suarez (1986 and 1987), Grossmann et al. (1988), Kaupenjohann and David (1996), Goyne et al. (2000)
sulfate (SO_4^{2-})	sorption	stainless steel, glass, oxide ceramic, Nylon, PE, PTFE, PVC			
zinc (Zn^{2+})	sorption, contamination	PA, PE, PTFE	stainless steel, oxide ceramic, PVC		Hädrich et al. (1977), Grossmann et al. (1985), Grossmann et al. (1990), Wenzel and Wieshammer (1995) Wenzel et al. (1997), Rais et al. (2006)

PA: polyamide (e.g. Nylon®), PE: polyethylen, PTFE: polytetrafluoroethylene (e.g. Teflon®)
glass: borosilica-glass (e.g. Duran®)
PVC: polyvinylchloride,

6.2.3.4. Sensor installation guide: required equipment, soil preparation, on-site installation steps

The installation of the suction cup into the soil is rather simple compared to other soil water sampling systems. In general, four installation modes are possible: i) horizontal with slight angle from a trench or pit, ii) vertical from the ground surface in slight angle, iii) vertical from the ground surface, and iv) vertical without shaft and backfilling of the auger hole (Mitchell et al. 2001) (Fig. 6.2).

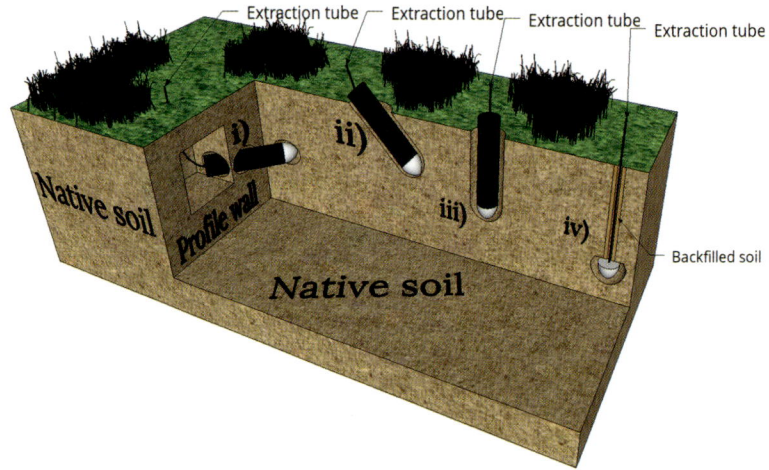

Fig. 6.2. Different possible installation methods for the suction cup. i) horizontal with slight angle from a trench or pit, ii) vertical with slight angle from ground surface, iii) from the ground surface, and iv) vertical without shaft and backfilling of the auger hole ground surface (drawing by Moritz Harings).

For the horizontal installation, an access trench has to be dug out, and the suction cups are installed into the native soil horizontally from the access trench. The extraction tubes have to be laid out to the top of the soil, where they can be connected to the sampling bottles. After the installation, the access trench will be refilled with the soil material (horizon by horizon). For the vertical non-shaft installation a hole will be drilled in the diameter of the suction cup and the porous cup will be installed into the soil. The rest of the auger hole will be again backfilled with native soil. Vertical and vertical in slight angle follow the same procedure with drilling to a predefined depths and installation of the suction cup. For these installations no backfilling is required, if the diameter of the soil auger fits perfectly to the diameter of the cup.

In all cases good hydraulic contact between the suction cup and the ambient soil should be guaranteed (Grossmann and Udluft 1991). Especially, in stony or coarse soils, a slurry made of the fine soil material collected from the respective depth or a suspension of quartz silt should be refilled into the borehole drilled for installation prior to suction cup installation to avoid gaps between natural soil and sampler.

For the installation, irrespectively of the installation from an access trench or from the soil surface, an auger with a diameter slightly smaller as the suction cup is required. In a first step, a hole at the position of the suction cups will be augered to the required depths and the auger material will be sampled in a bucket for the preparation of slurry or for backfilling. In a next step, the suction cup should be watered to facilitate installation into the soil. The end of the suction cups should be either coated by the slurry prepared from fine material of the auger samples or at sites with high skeleton or rock contents a slurry of artificial silty material should be used. At sites with extreme dry soils the auger hole should be wetted up to facilitate cup insertion. In a next step, the suction cup should be gently installed up to the end of the auger hole by turning the shaft. Hammering and abrupt for and back should be avoided to reduce breakdowns of the sensible cup material. After installation, the auger hole and/or access trench have to be backfilled. Extraction tubes have to be connected to sample bottles which should be located in a dark enclosure containing also the vacuum pump (if in use). For cold periods, the enclosures as well as all tubes have to be insulated or heated to avoid water freezing in the tubes and sample bottles.

6.2.4. Measurement and maintenance

Suction cups itself do not need any maintenance over the sampling period. Nevertheless, vacuum tests from time to time ensure proper functioning of the system and will detect breakdowns by e.g., feeding damages of mice. Sampling bottles should be cleaned or even better replaced at each percolation sampling to avoid algae growth.

6.2.5. Raw data processing and measurement error handling

A general problem often encountered is the calculation of mean concentration using several suction cups installed at one depth. It is generally known, that the mean concentrations from all samplers cannot be calculated directly by averaging the concentrations from individual samplers because each sampler will extract different amounts of soil water. Therefore, the correct way of calculating mean concentrations is to sum up all collected water and solute mass for one sampling period and calculate the mean concentration from the sum of extracted water and the sum of extracted mass.

Additionally, suction cup derived data (cumulative extracted water or solute concentration/mass) might vary substantially between samplers due to effects of soil heterogeneity. Especially, for the continuous operation mode with small offsets on matric potential sensor readings some samplers hardly extract water from the soil. These differences (especially low sampling rates) should be interpreted carefully and breakdowns of the system should be excluded before omitting such data.

The main problem interpreting suction cup data is that the influence of the suction cup on the soil water regime cannot be defined under natural flow conditions (Brandi-Dohrn et al. 1996, Warrick and Amoozegar-Fard 1977, Weihermüller et al. 2005, 2006, and 2011). Especially, the soil volume sampled and the imposed changes in matric potential on the natural flow pattern are not well known (Hart and Lowery 1997). It has also been speculated, that porous cups have an inherent bias in preferentially monitoring the chemical composition of larger soil pores at the expense of finer pores (Hansen and Harris 1975, Severson and Grigal 1976). Part of the confusion regarding the representativeness of the soil water extracted by suction cups probably results from the multitude of definitions used by different authors for the influence of the samplers on the soil water and extraction volume. To provide a consistent terminology for the interpretation and comparison of the behavior of suction cup in soils, Weihermüller et al. (2005 and 2011) introduced following characteristics:

1. The suction cup activity domain (SCAD), that represents the volume (cm^3) of influence in the matric potential distribution of the natural flow field after suction is applied to the cup.

2. The suction cup extraction domain (SCED), that defines the volume [cm^3] from which water and solutes can be extracted by a suction cup within a certain time interval.

3. The suction cup sampling area (SCSA), is defined as the area (cm^2) at the overlying soil surface from which water could be captured by the suction.

4. Leached mass fraction (LMF), which is the ratio of mass sampled by the mass applied at the soil surface multiplied by the infiltration surface divided by the SCSA.

Weihermüller et al. (2005) also showed that the SCAD, SCED, and SCSA depend on the soil hydraulic properties, the ambient water content, and the extraction time. As a consequence, Weihermüller et al. (2006) stated that the extraction domain and the sampling area are not constant in time and space for natural soils under atmospheric boundary conditions. This implies that the calculation of mass balances using porous cup data is not valid in its strict sense or is at least problematic without knowing the water flow within the soil (Weihermüller et al. 2011). Nevertheless, the LMF can be calculated using a water balance or hydrological model (Weihermüller et al. 2011).

Jury and Flühler (1992) suggested that suction cup samplers, particularly those installed at shallow positions, may be unreliable in soils with preferential flow, whereby recording of bypass flow might be a fundamental weakness of the cup sampler or a result of having too few samplers for adequate sampling of the flow regime. Therefore, the question of how many samplers are required remains, and whether it is practical to install and monitor the required number of suction cups (Flemming and Butters 1995, Weihermüller et al. 2006). Additionally, given the uncertainty regarding the SCED and the SCSA, the interpretation of sampled concentrations with suction cups is still a matter of debate.

In conclusion, suction cups are by far the most frequently used technique for extracting soil water. Easy installation and a large treasure of experience are the most important advantages of this technique. Especially, if solute transport and mass balances are in the focus of the experiment, the poor definition of the SCED and the SCSA together with the small cross-sectional area of the cup pose serious limitations on the interpretation of results. Furthermore, the necessity to use independent estimates of soil water fluxes increase the uncertainty of calculated fluxes. Alternative techniques, which aim at avoiding these limitations, have been proposed since the early beginnings of in-situ soil water extraction.

6.2.6. Related methods

- Field: suction plates, wick samplers, pan lysimeters, lysimeters, raisin boxes, matrix potential sensors
- Laboratory: suction plates, wick samplers, lysimeters, matrix potential sensors

6.3. Suction plates

The term suction plate or tension plate lysimeter is used for the single sampling device and also for arrays of such instruments. Sophisticated setups controlled by matric potential sensors and connected to additional control and measurement devices such as drop counters were often named differently such as 'multicompartment sampler' (Bloem et al. 2009 and 2010, Leinemann et al. 2016) or equilibrium tension plate lysimeter (ETPL) (Mertens et al. 2005). In the following the term suction plate will be used either for the single device as well as for arrays of suction plates. The principle of the suction plate is in some parts comparable to suction cups, and therefore, we will refer to the section describing suction cups (section 6.2) whenever possible to reduce redundancies.

Suction plates are widely used to monitor water and solute fluxes in the unsaturated zone, whereby the systems greatly differ in size and complexity. In general, suction plates consist of a flat porous body connected to an extraction tube (Fig. 6.3), which will be connected to sampling bottles. As for suction cups, various materials are in use for the plates such as porous ceramics, sintered glass or metal, or membranes (Weihermüller et al. 2007).

Suction plates can provide information of the water flux (mm or liter) and solute concentration (g l^{-1})

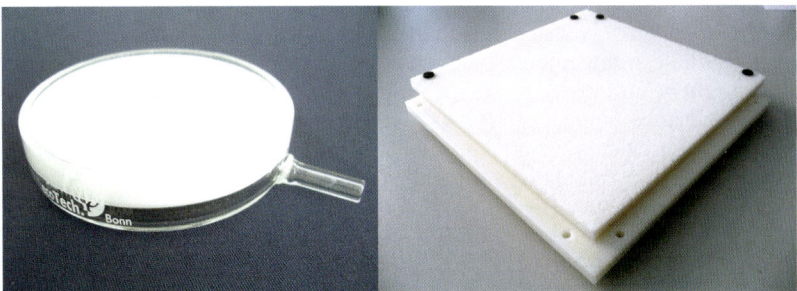

Fig. 6.3. Sintered glass plate (left) and porous nylon plate (right) (Photo: ecoTech GmbH, Bonn).

6.3.1. Operation of suction plates

Suction plates can be operated differently and various control options have been proposed. The simplest operation method is zero tension, where no suction will be applied to the suction plate. This operation mode has a great disadvantage because a saturated zone above the plate will be formed before water will enter the plate resulting in artefacts in water sampling such as divergent water flow away from the system, and therefore, underestimation of the natural water flux (Chiu and Shackelford 2000). Logically, not only water flow but also solute concentration in the sampled leachate will be biased by this seepage face boundary conditions as shown by Flury et al. (1999). In consequence, this operation mode is not recommended and instead of suction plates free drainage pan lysimeters should be used (see section 6.4).

The second operation mode is characterized by a fixed predefined suction applied to the plate. Correct results can be achieved with this method if a small and constant pressure is applied. Using –60 cm of water column (pF 1,8, field capacity), the plates theoretically collect exclusively soil water that is defined to be free draining by gravity through large pores (>50 µm). Sometimes, even a lower vacuum value is applied to the plates to get only preferential flowing water (Leinemann et al. 2016). With this set-up, the water coming from the imaginary column area above the plate is collected as seepage water. Under these preconditions, calculations of solute freights using the amount of water and the concentration (see section 6.2.5) are allowed. This is especially the case if those plates are installed in a depth below the root zone, where soil water is supposed to be only flowing downwards to the groundwater.

If the soil is drier than the above-mentioned value of –60 cm, the vacuum value has to be raised to extract soil water with suction plates. Unfortunately, the matric potential of the ambient soil generally varies as a function of space and time (see chapter 5). If soil water has to be collected using plates within those varying matric potentials, the soil water regime in direct vicinity of the plate is likely to be different from the fixed suction exerted by the plates. In consequence, the natural flow field of the soil water will be changed, and therefore, different soil water fluxes and solute concentration compared to the freely percolating water in the soil profile will be measured (Rhoades and Oster 1986, Kosugi and Katsuyama 2004).

To overcome the limitations described above, a suction equivalent to the ambient matric potential at the same depth of suction plate installation should be applied to the suction plate to representatively sample soil water and solute concentrations (van Grinsven et al. 1988, Byre et al. 1999 and 2001, Foley et al. 2003, Lentz and Kincaid 2003, Pelger et al. 2003, Siemens et al. 2003, Barzegar et al. 2004, Kosugi and Katsuyama 2004, Siemens and Kaupenjohann 2004, Mertens et al. 2005). The ambient matric potential is generally measured by reference matric potential sensors (see chapter 5) and automatically applied to the plates. This control strategy is expected to sample representative water flow and solute concentrations with little disturbance of the natural flow field. As already stated in section 6.2.1, most available systems use only one reference sensor for a whole arrangement of suction devices, and therefore, an unknown error of the ambient suction might be introduced by soil heterogeneity. In consequence, these systems are not only cost- and maintenance-intensive, but also give the impression of having a high-sophisticated technical solution for calculating solute freights. Hence, resulting solute concentrations and solute transport parameters calculated from solute breakthrough curves sampled with suction plates should be verified using water balance models, water tracers, or numerical simulations (Siemens et al. 2003, Kosugi and Katsuyama 2004, Siemens and Kaupenjohann 2004, Mertens et al. 2005, Kasteel et al. 2006).

6.3.2. Continuous data recording or single observations

As already described for the suction cups (see section 6.2) suction plates integrate over a certain time period, and therefore, data with high resolution are rare. Also, the amount of extracted water should be sufficiently large for chemical analysis. Sampling intervals (exchange of sampling bottles) depend on the hydraulic conditions of the site, whereby sampling and exchange of sample bottle (extract) should be in smaller time intervals during periods of high leaching (e.g., wet periods) and can be longer during drier conditions. In all cases overflow of the sampling bottles should be avoided.

6.3.2.1. Range of application

The range of application for suction plates is comparable to suction cups, and therefore, we refer for general applications to section 6.2. The advantage of suction plates over suction cups is that solute freights can be better estimated due to the 2-dimensional character of the sampler, which reduces bypass flow.

6.3.2.2. On-site infrastructure and other measurements

The on-site infrastructure necessary for operating suction plates is comparable with those necessary for operating suction cups with matric potential control (suction control) (section 6.2.2). Because the operation mode without application of suction is not recommended, either a special constant vacuum regime should be used, or, in case of matric potential driven vacuum, additional sensors such as matric potential sensors as well as a control and data storage unit are required.

6.3.2.3. Soil type: Depth of interest, soil texture, obstacles: roots, skeleton

As for suction cups, suction plates can be installed and operated at all locations and soil types, whereby the installation depth is only limited by the accessibility (e.g., depth of access trench). For more details we refer to the section 6.2.2.1.

6.3.2.4. Other constrains

Because suction plates and suction cups are generally of same type of in-situ sampler, we refer to section 6.2.2.2 for general details. Because installation of suction plates is far more complex as the installation of e.g., suction cups and often suction plates are higher in price, these additional points should be taken into account in the funding proposal. Additionally, the recommendations listed in section 6.2 should be taken into account also for suction plates.

6.3.3. Implementation of suction plates

6.3.3.1. Installation of suction plates

For the installation of suction plates an access trench has to be dug out, and the suction plates are installed into the native soil horizontally from the access trench. Alternatively, suction plates can be installed from the soil surface by removing the top soil layer, adding in the plate and refill the dig out afterwards.

Because the native soil will be destroyed by this procedure, this installation is mainly restricted to the plough horizon or to installations where a re-installation of a cut-out soil layer is possible (see Fig. 6.4a). It is highly important to ensure that the porous plate has good hydraulic contact to the surrounding soil. Therefore, plates installed from an access trench have to be installed in horizontal tunnels dug into the native soil. Ceramic, glass, plastic, or metal plates should be watered prior to keep the slurry wet. In these tunnels a support frame can be build-in, whereby the support frame not only holds the porous plate but should also allow the pressing of the plate to the top of the tunnel by e.g., spindles. Alternatively, the space beneath the suction plate has to be refilled very carefully with the original soil material. To avoid loss of contact due to setting effects in this refilled material, it has to be compressed carefully by hammering on a square timber or similar to reach a bulk density at least as high as the surrounding material. Be sure that the plate itself is not affected by this procedure.

In comparison to suction cups, not only in stony or coarse soils, a slurry made of the fine soil material collected from the respective depth or a suspension of quartz silt must be placed on top of the plate prior installation to avoid gaps between the natural soil and sampler (see Fig. 6.4b). The reason to add the slurry is that the top of the access tunnel which acts as the contact area will never be completely flat. After installation, the extraction tubes have to be laid out to the top of the soil or to a permanent manhole, where they can be connected to the sampling bottles and control unit. After the installation the access trench will be refilled with the soil material (horizon by horizon).

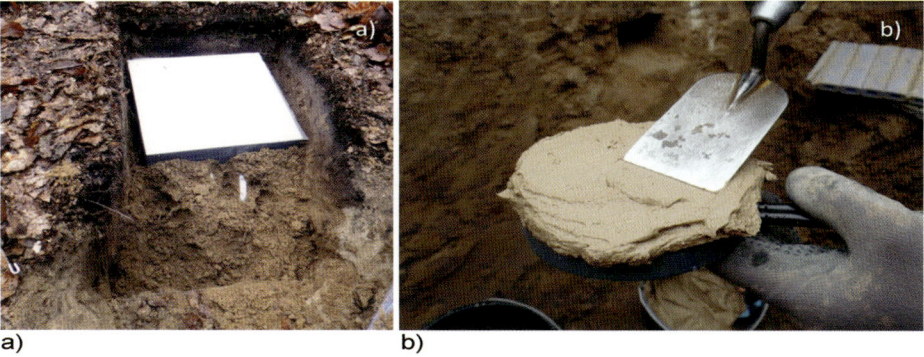

Fig. 6.4. a) Installation of a suction plate from the surface (Photo: Group Leinemann et al.); b) Adding slurry on the plate surface prior installation (Photo: ecoTech GmbH, Bonn).

6.3.2.2. Experimental layout: horizontal and vertical sensor allocation, spatial replicates, spatial representativity of point observations

Installation of suction plates is mainly determined by the boundary conditions of the experiment (e.g., accessibility). Suction plates can be also installed at or close to horizon boarders because sampling will mainly come from the overlying soil. In comparison to suction cups, suction plates exhibit a larger sampling area. Therefore, the detection of preferential flow events may be possible using suction plate arrays (Ciglasch et al. 2005, Leinemann et al. 2016). Due to the 2D surface of the plates, the origin of the sampled water and solutes is also better defined as for suction cups, which supports also mass balance estimations (Kasteel et al. 2006). These advantages, however, are associated with larger efforts and disturbance of experimental plots for installation as compared to the installation of suction cups.

6.3.2.3. Self-construction of suction plates

Theoretically, suction plates can be self-constructed, whereby the most critical part of the sampler are the porous materials, which normally do have a large air entry value, and therefore, require special materials not always available at the free market. Additionally, material should be free of contamination – this should include tubing's and glues.

6.3.2.4. Sampler calibration

As suction cups suction plates cannot be calibrated but some general testing should be performed prior installation. For details we refer to section 6.2.3.3.

6.3.3. Measurement and sensor maintenance

Suction plates itself do not need any maintenance over the sampling period. Nevertheless, vacuum test from time to time ensure proper functioning of the system and will detect breakdowns by e.g., feeding damages of mice. Sampling bottles should be cleaned or even better replaced at each percolation sampling to avoid algae growth.

6.3.4. Raw data processing and measurement error handling

As described in section 6.2.5 averaging solute concentration over several suction plates is often not done in the right way.

Additionally, suction plate derived data (cumulative extracted water or solute concentration/mass) might vary between samplers due to effects of soil heterogeneity as shown by Kasteel et al. (2006). Also upward flow during dry conditions with corresponding downward flow after this period might artifact the results obtained by suction plates (Kasteel et al. 2006). Therefore, numerical modelling or the use of water balance models will help in the interpretation of the measured data and to get knowledge of potential processes involved.

6.3.5. Related methods

- Field: suction cups, wick samplers, pan samplers, lysimeters, raisin boxes, matrix potential sensors
- Laboratory: suction cups, wick samplers, lysimeters, matrix potential sensors

6.4. Pan lysimeters

Pan or zero-tension lysimeters are passive samplers in the shape of a pan without large side walls extending above the system (Fig. 6.5). In the following, only the term pan lysimeter will be used to avoid confusion with the classical zero-tension lysimeters discussed in section 6.6. Pan lysimeters collect freely percolating soil water only, which distinguishes pan lysimeter from most other sampling devices (Ebermayer 1873, Jemison and Fox 1992). The pan lysimeter system itself can be made of different materials such as steel, stainless steel, glass, ceramic, or plastic material depending on the scientific question and the target substance. Pan lysimeters can provide information of the water flux (mm or liter) and solute concentration (g l^{-1}).

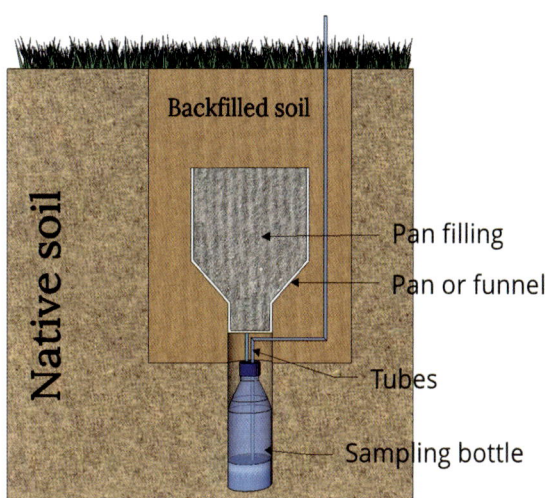

Fig. 6.5. Schematic of a pan lysimeter installed in the field (drawing by Moritz Harings).

6.4.1. Operation of pan lysimeters

The operation of pan lysimeters is comparably easy because the system is passive and water and solutes will be collected without any suction applied to the system or any electronically devices needed. The only time expensive part in the operation is the exchange of the sampling bottles (if accessible from e.g., a permanent manhole) or the pumping of the leachate from buried bottles to the soil surface.

6.4.1.1. Continuous data recording or single observations

As already mentioned, pan lysimeters are passive systems and leachate will only be sampled during relatively wet soil conditions. Therefore, permanent data series are often not available and measured water fluxes as well as solute concentrations/masses might not fully reflect the natural system.

6.4.1.2. Range of application

In general, pan lysimeters are placed below the ground surface to capture drainage water, whereby the sampling surfaces of the system can exhibit several square meters (e.g., in waste deposal sealing's) with standard dimensions being about 0.5 m². Therefore, some of the natural heterogeneity in water and solute flux will be captured as well. Pan lysimeters can be also used in shallow depth, e.g., in organic or forest litter layers and then are called humus lysimeters (Marques et al. 1996, Ranger et al. 2001). For soil in dry conditions pan lysimeters are less suitable because water flow will rarely cause local saturation above the pan lysimeter system, which is a prerequisite for water sampling. On the other hand, pan lysimeters are the cheapest and easiest soil water extraction system and need only low maintenance, and therefore, are a good alternative for screening applications.

6.4.1.3. On-site infrastructure and other measurements

Due to the passive sampling character of the system no on-site infrastructure is needed. On the other hand, additional information about the water flow (e.g., soil water sensors, see chapter 4) are desirable for better interpretation of the extracted water volumes and solute concentrations.

6.4.1.4. Soil type: Depth of interest, soil texture, obstacles: roots, skeleton

Based on the design of the sampler and the absence of capillary connection to the ambient soil, pan lysimeters operate reasonably well in soils with large macropores near saturation but are much less successful if the soil dries out (Zhu et al. 2002). Especially, for organic rich soils as well as litter layers which often do not allow the installation of other samplers requiring full capillary contact, pan lysimeters are beneficial.

6.4.1.5. Other constrains

Pan lysimeters are by far the cheapest soil water extraction device, and therefore, larger numbers of replicates can be installed and operated. Especially, in remote location where no permanent power supply (including batteries) is available or allowed pan lysimeters can be used. On the other hand, the clear disadvantages of the system have to be taken into account for decision making. Additionally, the recommendations listed in section 6.2 should be taken into account also for pan lysimeters.

6.4.2. Implementation of pan lysimeters

6.4.2.1. Installation of pan lysimeters

The installation of the pan lysimeter requires a filling with coarse gravel or some other highly water conductive materials to guarantee easy interception of the drainage water and to divert it to the collection device. Placing a gravel drain in the soil subsurface generally creates a seepage flow boundary condition with a pressure head equal to atmospheric pressure (Richards 1950). Therefore, the soil saturates at the interface between the natural soil and the gravel filling. If the pan is filled with coarser material as the ambient soil, a tendency for water bypass is imposed as a response of the water potential gradients existing in the soil at the interface and the soil surrounding the system. The amount of bypass flow strongly depends on the water flux rate, the textural contrast between the filling material and the surrounding soil, and the gradients in water potential that persist in and around the pan lysimeter.

For the field installation, the pan lysimeters will be either installed from the surface or from an access trench as already described for suction cups or suction plates (see section 6.2.3.4 and 6.3.3.1).

6.4.2.2. Experimental layout: horizontal and vertical sensor allocation, spatial replicates, spatial representativity of point observations

The number of replicates need for capturing the natural heterogeneity in water and solute flux greatly depend on the size of the pan lysimeter. Large scale pan lysimeters need generally less replicates, whereas small scale pan lysimeters should be installed in higher numbers. With respect to the price high numbers of replicates are easily achievable. Because pan lysimeters are often used to capture preferential flow, larger numbers of samplers or large samplers should be used.

6.4.2.3. Self-construction of pan lysimeters

Pan lysimeters are easy self-constructed, whereby the pan itself can be made out of different materials (glass, ceramic, metal, plastic). The choice of material greatly depends on the size of the pan lysimeter and the substances under investigation (see Table 6.1). For the filling of the pan lysimeter, commercially available products (sand and gravel) can be used. All filling materials should be washed before filling to avoid cross-contamination by the filling material.

6.4.2.4. Sampler calibration

Pan lysimeters cannot be calibrated and the functionality in the field cannot be tested easily. Even if water will be sampled by pan lysimeters large amounts of percolate might bypass the sampler. Therefore, it is recommended to apply conservative tracers (chloride or bromide) on the surface above the sampler and calculate the mass balance of sampled leachate to estimate the bypass flow.

6.4.2.5. Sensor installation guide: required equipment, soil preparation, on-site installation steps

The installation of pan lysimeters is comparable to the installation of suction plates (see section 6.3.3.1), whereby less effort has to be made to ensure hydraulic contact between the native soil and the pan lysimeter.

6.4.3. Measurement and sensor maintenance

Pan lysimeters itself do not need any maintenance over the sampling period but sampling bottles should be cleaned or even better replaced at each percolation sampling to avoid algae growth, especially if the pan lysimeters are installed in organic rich horizons (e.g., litter), where large amounts dissolved organic carbon (DOC) and particulate organic matter (POM) are in the drainage water.

6.4.4. Raw data processing and measurement error handling

Initially, pan lysimeters were used primarily to analyze water quality and only occasionally to quantify drainage rates. More recently, zero-tension lysimeters have been used to estimate drainage rates over a wide range of soil conditions (Chiu and Shackelford 2000, Zhu et al. 2002, van der Velde et al. 2003, 2004). Because of water divergence, collection efficiencies less than 10% have been noted for pan lysimeters (Jemison and Fox 1992, Zhu et al. 2002). Therefore, diversion of water around zero-tension lysimeters can be a significant problem. Flury et al. (1999) also showed in numerical simulations that these seepage face conditions not only influence the water

116 Soil water parameters

flow, but also the solute concentration in the sampled leachate. Disadvantageous are their complex installation, that causes considerable disturbance on experimental plots and the divergence of water flow around the system. This prevents quantitative estimates of flux concentrations, and therefore, complicates the interpretation of solute breakthrough and which may even lead to complete failure of the system.

Humus lysimeters show less problems with saturation and bypass flow, due to the fact, that humus has a more coarse and open-pored structure. Humus lysimeters normally only have a nylon mesh at the top and are not filled with gravel or other mineral materials, because the water flowing out of the humus layer should not have contact to mineral surfaces which would cause flocculation and/or changes in solute chemistry (Guggenberger and Zech 1993). For averaging measured concentration from different samplers, the recommendations provided in section 6.2.5 should be followed.

6.4.5. Related methods

- Field: suction plates, suction cups; wick samplers, lysimeters, raisin boxes; soil water content measurements

6.5. Capillary wicks

Capillary wick samplers or wick lysimeters are sampling devices which sample soil water by the gravitational potential using an inert wick material such as fiberglass (Holder et al. 1991) or rock wool (Ben-Gal and Shani 2002) (Fig. 6.6). In the following, only the term capillary wick will be used for the entire sampling device. By applying a hanging water column using wicks, the drainage water is pulled out of the sampling device, while the lower soil boundary is maintained at a pressure less than atmospheric, resulting in an unsaturated soil. Wick samplers are easy to build up and do not need any electrical power supply to generate suction for water extraction, and are therefore, a perfect sampling device in remote locations and for low budget instrumentation.

Capillary wicks cups can provide information of the water flux (mm or liter) and solute concentration (g l^{-1}).

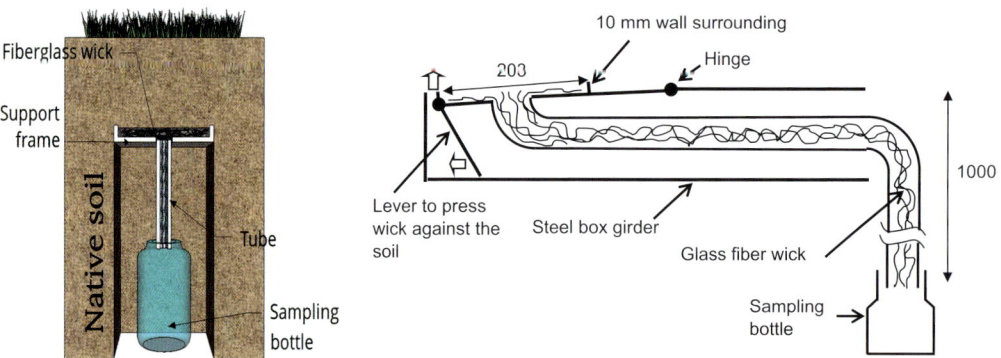

Fig. 6.6. Schematic of an installed capillary wick sampler (drawing left figure by Moritz Harings).

6.5.1. Operation of capillary wicks

As mentioned, capillary wick samplers sample soil water and solutes by the gravitational potential using an inert wick material. The degree of suction which will be imposed by the wick depends upon the material, length, and diameter of the wick, the dimension of the sampling bottle, the flux rate, and the soil type (Holder et al. 1991, Boll et al. 1992, Knutson and Selker 1994, Rimmer et al. 1995, Zhu et al. 2002, Mertens et al. 2007). The maximum suction which can be generated by the hanging water column inside of the wick is about 50–100 cm (Holder et al. 1991, Boll et al. 1992, Brandi-Dohrn et al. 1996). For wick-type lysimeters, drainage bypass can be further minimized by placing an extension tube above the wick (Gee et al. 2002, 2003). The extension tube is filled with soil from the excavation of the hole into which the wick sampler is placed. If the properties of the wick-sampler are adjusted to the soil properties, a semiqualitative, direct analysis of the water and solute flow is possible (Boll et al. 1992, Knutson and Selker 1994, Rimmer et al. 1995, Louie et al. 2000, Zhu et al. 2002, Siemens and Kaupenjohann 2004).

6.5.2.1. Continuous data recording or single observations

As suction cups, suction plates, and pan lysimeters, capillary wicks generally integrate over a certain time period mainly defined by the exchange of the sampling bottles (see section 6.2). To increase the temporal resolution at least for the water flux Gee et al. (2002 and 2003) integrated a tipping bucket and Mertens et al. (2008) an automated low-cost drop counter into their wick sampler. Theoretically, both systems (tipping bucket or drop counter) can be also implemented into suction cups and plate systems, whereby due to the active suction applied to both systems more effort is needed for the integration.

6.5.2.2. Range of application

Capillary wicks can be used for a wide range of applications in the field and also laboratory, whereby the extraction of soil water is limited to the pressure generated by the maximum length of the hanging water column (max ~100 cm). Therefore, only relatively wet soils will be sampled. Especially, for coarse soils (e.g., sand) this threshold can be easily reached at dry soil conditions or if the sampler is installed close to the soil surface.

For most chemicals in the soil solution no chemical alteration or changes of the transport parameters (dispersion and retardation) were determined by the used wick material (Holder et al. 1991, Knutson and Selker 1994, Siemens and Kaupenjohann 2004). In contrast, findings by Goyne et al. (2000) showed that fiber-glass-wicks alterated pH, alkalinity, calcium, magnesium, potassium, aluminum, and silicon concentrations in the sampled percolate of acid forest soils by weathering of the wick material.

Nevertheless, capillary wick samplers seem to offer a compromise between rather costly matric potential controlled suction plate systems and pan lysimeters.

6.5.2.3. On-site infrastructure and other measurements

As already mentioned, capillary wick systems impose a pressure by the hanging water column, and therefore, no additional on-site infrastructure such as power supply is needed. On the other hand, additional information about the water flow (e.g., soil water sensors; see chapter 4) are desirable for better interpretation of the extracted water volumes and solute concentrations, especially if the sampler will not extract water over the entire measurement period due to drying out of the soil.

6.5.2.4. Soil type: Depth of interest, soil texture, obstacles: roots, skeleton

In general, capillary wick samplers can be installed and operated at all locations and soil types. The installation depth is only limited by the accessibility (e.g., depth of access trench). For soil rich in skeleton or rocks special care is needed while installation of the wick samplers to ensure good hydraulic contact between the soil and the wick material. For dry soil conditions (e.g., in arid or semiarid zones) it has to be checked if the pressure imposed by the hanging water column is large enough to extract water from the soil at least for the drainage periods.

6.5.2.5. Other constrains

Capillary wick samplers are quite cheap soil water extraction devices, and therefore, larger numbers of replicates can be installed and operated. Especially, in remote location where no permanent power supply (including batteries) is available or allowed, capillary wick samplers can be used. On the other hand, the clear disadvantages of the system have to be taken into account for decision making. Additionally, the recommendations listed in section 6.2 should be taken into account also for capillary wick samplers.

6.5.3. Implementation of capillary wicks

6.5.3.1. Installation of capillary wicks

For the field installation, capillary wick systems will be either installed from the surface or from an access trench as already described for suction cups or suction plates (see section 6.2.3.4 and 6.3.3.1)

6.5.3.2. Experimental layout: horizontal and vertical sensor allocation, spatial replicates, spatial representativity of point observations

The experimental layout for experiments using capillary wick systems is comparable to those using suction plates. Therefore, we refer the reader to section 6.3. With respect to the price of capillary wick systems, high numbers of replicates are easily achievable. Because capillary wick systems are often used to capture preferential flow events, larger numbers of samplers or large samplers should be used.

6.5.3.3. Self-construction of capillary wick samplers

Capillary wick samplers are not commercially available and thus have to be self-constructed. In general, the capillary wick system consists of the wick, which is commercially available, and a support frame in which the wick will be laid out (see Fig. 6.6). The wick below the support frame will be inserted into a plastic (e.g., PTFE) tube to prevent drying out of the wick. The end of the wick will be inserted into the sample bottle. It has to be ensured that the wick hangs freely below the support frame. The pressure applied to the soil equals the distance between the support frame and the wick-end.

6.5.3.4. Sampler calibration

It has to be noted, however, that the wick materials used for construction of the samplers are produced in large scale industrial processes for e.g., heat insulation purposes. This means that variations of their hydraulic and chemical properties are likely to be large with respect to their use as scientific equipment. This in turn means that substantial testing of hydraulic, chemical, and sorptive properties of the samplers in pre-experiments is inevitable. It has been also recommended to anneal the wick material before installation (400 °C for 12 h) to remove potential contaminations by oils and finishing (Knutson and Selker 1994). After that, the wick should be percolated with soil solution to test potential sorption and release of the target substances.

The still rather small amount of operating experience gained with this technique also requires a sound testing of samplers prior to their large-scale use in experiments. This could be e.g., the use of conservative tracers (chloride or bromide) applied on the soil surface above the samplers to analyze the sampling capacity of the capillary wick samplers.

6.5.3.5. Sensor installation guide: required equipment, soil preparation, on-site installation steps

The installation of capillary wick systems is comparable to the installation of suction plates (see section 6.3.3.1). Because the wick has to hang below the support frame, wick samplers generally are installed from a permanent access, e.g., manhole. Fully buried wick sampler systems are also possible such as the 'passive wick fluxmeter' described by Gee et al. (2009).

To prepare the installation of capillary wick systems, Mertens et al. (2007) used numerical simulations, whereby these simulations were used to optimize the wick sampler design.

6.5.4. Measurement and sensor maintenance

Capillary wick samplers itself do not need any maintenance over the sampling period. Sampling bottles should be cleaned or even better replaced at each percolation sampling to avoid algae growth.

6.5.5. Raw data processing and measurement error handling

For averaging measured concentration from different capillary wick samplers, the recommendations provided in section 6.2.5 should be followed. The problem of representativity of the measured data is comparable to those discussed in section 6.3.4 for suction plates.

Because capillary wick samplers do not extract water during dry periods (or at periods where the matric potential exceeds the pressure imposed by the hanging wick), data should be interpreted carefully. In extensive field testing over several years, leachate collection efficiencies (LCEs), defined as measured drainage divided by estimated drainage (obtained from a mass balance of precipitation and evapotranspiration), have been shown to equal or exceed 100% for wick samplers (Louie et al. 2000, Zhu et al. 2002), whereby these findings might be restricted to the site of installation.

6.5.6. Related Methods

- Field: suction plates, suction cups, pan lysimeters, lysimeters, raisin boxes, matrix potential sensors
- Laboratory: suction plates, suction plates, lysimeters, matrix potential sensors

6.6. Lysimeters

Lysimeters are containers or vessels containing disturbed or undisturbed soils. Often the term soil column is used for the same system but the term soil column is mainly restricted to lysimeters smaller in size and used in laboratory setups.

The general aim of lysimeter studies is the measurement of the components of the water balance, namely precipitation, actual evapotranspiration, changes in the water storage, and/or drainage, whereby simple lysimeters are restricted to measure water drainage only. For the measurements of all components of the water balance, especially actual evapotranspiration and precipitation, the lysimeter has to be weighable (Peters et al. 2014). Lysimeters have been used in agricultural studies to measure ground water recharge (Yang et al. 2000), solute transport towards the groundwater (Schoen et al. 1999, Kasteel et al. 2010), water fluxes at the soil-plant-atmosphere interface (e.g., Meissner et al. 2007), as well as in urban sites to study surface runoff (Nehls et al. 2011). Additionally, lysimeters with groundwater control are used to measure the soil water balance of sites influenced by groundwater (Bethge-Steffens et al. 2004). If the lysimeter is equipped for volatilization measurements, a closed mass balance can be calculated even for volatile substances (Führ et al. 1998, Meissner et al. 2000, Wolters et al. 2003). A general picture of a lysimeter setup is shown in Fig. 6.7.

Irrespectively of the choice of system setup (weighable or not), drainage will be monitored in liter or mm, and concentration of substances in the extracted water in g l^{-1}. Weighable lysimeters allow also the estimation of actual evapotranspiration and precipitation in liter or mm.

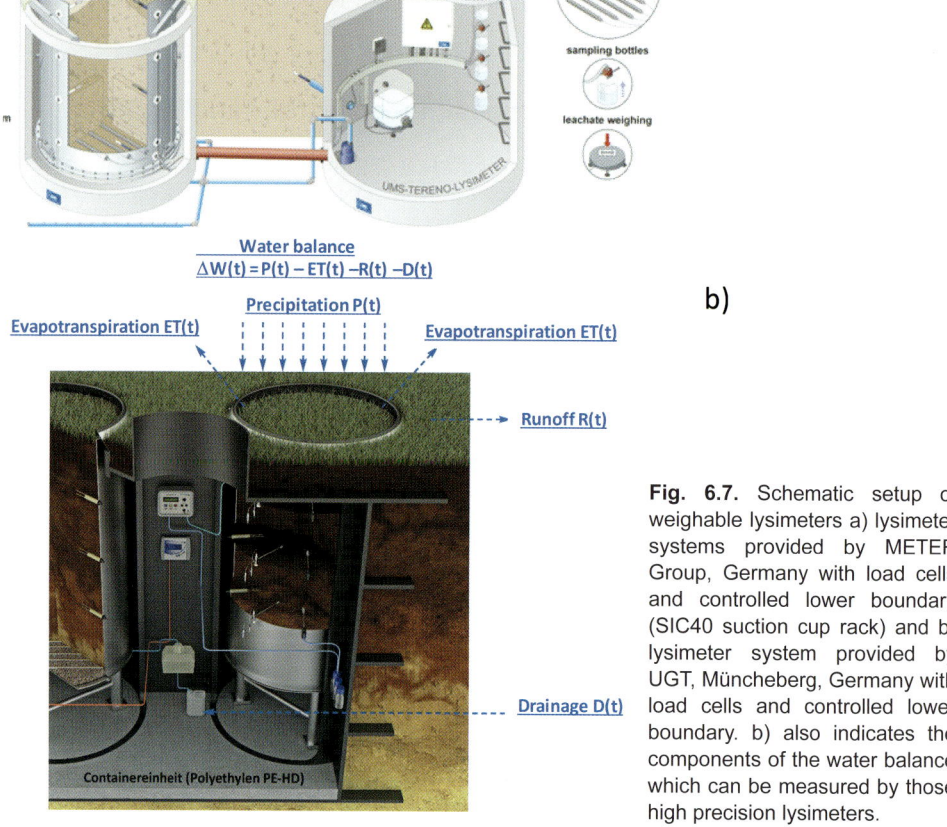

Fig. 6.7. Schematic setup of weighable lysimeters a) lysimeter systems provided by METER Group, Germany with load cells and controlled lower boundary (SIC40 suction cup rack) and b) lysimeter system provided by UGT, Müncheberg, Germany with load cells and controlled lower boundary. b) also indicates the components of the water balance which can be measured by those high precision lysimeters.

6.6.1. Operation of lysimeters

By the drainage behavior of water from the system, two types of lysimeters can be distinguished: i) Free drainage lysimeters, where water is allowed to drain freely through the soil under gravity, or ii) suction controlled drainage system, where a defined suction is imposed at the lower boundary using suction cups, wick samplers, or porous plates.

In general, a free drainage lysimeter is easier to install and cheaper than the controlled suction system. A major concern of the free drainage lysimeter is that the lower boundary is exposed to the atmospheric pressure, resulting in an evolution of a water saturated zone at the bottom of the lysimeter before drainage (Abdou and Flury 2004). This lower boundary can impose temporary anaerobic conditions, which may influence degradation, solute transport, and changes in the capillary rise during evapotranspiration (Bergström 1990, Giessler et al. 1996). In comparison, suction lysimeters are more expensive and difficult to install, especially if they have large surface areas (Bergström 1990). Another problem with suction controlled lysimeters is that water and solutes can interact with the material used for the suction devices (see Table 6.1). Additionally, the natural matric potential, water flow streamlines, and the composition of the leachate can be altered. The drainage patterns of both systems have been compared in several laboratory and numerical experiments, with the general finding that suction lysimeters drain more water continuously and in larger quantities (Colman 1946, Dowdell and Webster 1980, Vereecken and Dust 1998, Abdou and Flury 2004). The latest generation of lysimeters combine the control of the lower boundary (suction control) with additional matric potential sensor measurements to mimic an 'endless' soil column. These reference sensors are installed in the undisturbed soil in close vicinity of the lysimeter. In this case, the suction at the lower end of the lysimeter equals the matric potential in the undisturbed soil, and water can be either extracted from the lysimeter or pumped into it when capillary rise or water table fluctuations will cause upward positive water fluxes. Additionally, weighing resolution and weighing precision have been substantially improved, and therefore, modern lysimeters resting on load cells can reach resolution of up to 0.01 mm fluxes (drainage, evapotranspiration, or precipitation) (von Unold and Fank 2008). These type of lysimeters are currently regarded as the most precise and expensive measurement systems for rainfall, actual evapotranspiration, or even dewfall (Meissner et al. 2007)

In general, operation of a lysimeter itself is fairly easy because only the weighing data have to be stored (if the lysimeter is weighable) and the drainage water has to be collected. Nevertheless, additional measurement devices often installed into the lysimeter such as water content sensors (chapter 4), temperature probes, suction cups (chapter 6.2), or matric potential sensors (chapter 5) need maintenance and data interpretation.

6.6.1.1. Continuous data recording or single observations

The difference in continuous data recording or single observations is mainly restricted by the lysimeter system itself. If the system is weighable, continuous data recording is normally provided, whereby for the quantification of precipitation and actual evapotranspiration reasonable small-time measuring intervals are required (range of minutes) (Ramier et al. 2004). For simple non-weighable systems, the resolution is much coarser depending on sampling intervals of the drainage water, but here, only the amounts of drainage water and containing solutes are of major interest.

6.6.1.2. Range of application

As mentioned above, lysimeter have been used for a wide range of applications ranging from agricultural to soil-plant-atmosphere studies. There are also no limits in the applicability in terms of soil used in lysimetry. For soils with extreme skeleton content or where the underlying bedrock has to be studied, the undisturbed extraction of the lysimeter is the only limitation. Also, the size of the lysimeter might limit some applications. For example, even forest lysimeters of the size of several 10 of meters are reported, whereby these lysimeters are generally not weighable. The classical weighable lysimeter size ranges from several dm in length and diameter to about 3 m in length and 2 m in diameter.

6.6.1.3. On-site infrastructure and other measurements

The on-site infrastructure necessary for operating lysimeters greatly depends on the choice of the lysimeter system. For simple free drainage lysimeters no power supply is needed, and the lysimeter vessel can directly be installed in the field. To get information about the atmospheric boundary conditions, which are essential to gain information about potential evapotranspiration and precipitation (necessary for the calculation of the water balance), a climatic station is additionally needed. If sensors will be installed into the free drainage lysimeters (e.g., water content or matric potential sensors), the infrastructure necessary for the operation of this equipment is also needed.

For the operation of weighable and weighable suction-controlled lysimeters, the on-site infrastructure has to be much more sophisticated ranging from permanent power supply, data storage capacities, up to data transfer. The setup often requires also a central service manhole, where the data-logger, pumps, drainage water storage tanks etc. are installed and the lysimeter can be maintained.

6.6.1.4. Soil type: Depth of interest, soil texture, obstacles: roots, skeleton

The optimal surface area and the lysimeter length depend mainly on the scientific question, the filling procedure, the lower boundary, and the location of installation. The base area is strongly connected to the scale of observation, whereby small-scale heterogeneity will be averaged using large base areas. Lysimeters with crop stands should represent the natural crop inventory and the maximal root depth should be taken into account. In general, lysimeters can be filled either monolithic or with disturbed soils.

A major concern of the lysimeter concept is that it does not account for lateral water and solute fluxes, and that the vertical boundaries may cause fringe effects and preferential flow paths (Schoen et al. 1999). To reduce such fringe effects, care should be taken in the lysimeter extraction or filling with disturbed soils.

6.6.1.5. Other constrains

As for most soil water extraction or observation systems it should be considered that water and solute movement in the soil is a rather slow process and deep percolation might take several months to even years. This holds especially for reactive (sorbing) substances. To get a rough estimate of the minimal time period required, hydrological and solute transport modeling might help. The choice of the lysimeter type (free drainage, weighable, or weighable/suction controlled) greatly depends on the experimentation length and funding sources. For short, observation periods with low fundings, free drainage lysimeters might be preferable as long as the system limitations are acceptable.

6.6.2. Implementation of lysimeters

6.6.2.1. Installation of lysimeters

The installation effort of lysimeters also depends on the choice of the lysimeter system. Simple small to medium sized free drainage and weighable lysimeters can be relatively easily filled monolithic by pressing the lysimeter vessel into the soil e.g., by the help of a power shovel. Restrictions in this procedure are that the soil must not be rich in skeleton and has to be deep enough. Larger lysimeters as well as the extraction of soils rich in skeleton need special equipment which is provided by some companies selling complete lysimeter solutions.

Lysimeters which will be filled non-monolithic are much easier to fill but care should be taken to fill it layer by layer with a certain range of compaction for each soil layer.

Because lysimeters and especially lysimeter clusters often need excavations and construction works, special companies should be hired for this purpose.

6.6.2.2. Experimental layout: horizontal and vertical sensor allocation, spatial replicates, spatial representativity of point observations

As already mentioned, lysimeters are much more representative compared to other in-situ water extraction samplers, due to the generally larger base area. Nevertheless, one lysimeter itself rarely represents the field water and solute flow, as nicely shown by Kasteel et al. (2007 and 2010). Representativity increases with base area, and suction controlled weighable lysimeters are the best choice for reliable estimation of all components of the water balance and suction controlled weighable lysimeters represent the natural field conditions most.

6.6.2.3. Self-construction of lysimeters

Theoretically, lysimeters can be self-constructed, especially small to medium scale free drainage lysimeters. This self-construction starts by using commercially available pipes of suitable length and diameter and stops at the self-construction of complete lysimeter stations and clusters. Care should be taken in the material choice of the lysimeter (see Table 6.1)

6.6.2.4. Sampler calibration

Lysimeters itself cannot be calibrated. Nevertheless, calibration might be necessary for load cells and additional sensors used in the lysimeter experiment.

Lysimeters should be also extracted or filled several weeks or month before the start of the experiment to allow setting of the soil filled in (for non-monolithic lysimeters). Also, monolithically filled lysimeters should not be extracted right before the start of the experiment to allow the soil to relax. This is necessary because stress can be imposed during the monolithic filling, especially close to the side walls.

6.6.2.5. Sensor installation guide: required equipment, soil preparation, on-site installation steps

For the installation of the lysimeters, the upper surface should be equal to the top ground surface to minimize microclimatic changes. Additionally, the space between the lysimeter vessel and the cavity where the vessel will be placed in (for weighable lysimeters) should be also minimized to reduce artificial temperature gradients within the soil block. If possible, these lysimeters should be insulated at the vertical sites to guarantee a horizontally layered temperature profile in the soil. The lower boundary can be segmented to obtain information on the spatial heterogeneity of the water and solute fluxes (Schoen et al. 1999). For acquisition of the surface run-off, the lysimeters can also be equipped with run-off/overflow tubes (van Weesenbeck et al. 1998).

Because the installation of large weighable (suction controlled) lysimeters is much more difficult, specialized companies are required.

6.6.3. Measurement and sensor maintenance

Lysimeters generally do not need any special maintenance over the experimental period, and only drainage water has to be collected at certain intervals. If sophisticated weighable (and suction controlled) lysimeters will be used, maintenance increases substantially because the additional pumps, load cells, etc. have to be maintained. This also holds for any other sensors installed into the lysimeter.

6.6.4. Raw data processing and measurement error handling

If only drainage water will be analyzed, e.g., from free drainage lysimeters, the data processing of lysimeter data is fairly simple because only the amount of seepage water has to be registered. In this mentioned case, the water volume sampled have to be related to the base area per unit time (sampling interval). Additionally, measured concentrations will be handled in the same way. For weighable lysimeters, the data processing is far more complicated because larger amounts of data are available and the raw data from the balance or load cells might be affected by wind effects and maintenance. Also, the extraction of water from additional samplers such as installed suction cups must be corrected. Especially, wind effects might cause some noise in the raw weight readings, so they cannot be straightforward interpreted as a weight increase (precipitation) or loss (evapotranspiration or drainage). To get reliable weight results, the noise has to be filtered out (Fank 2013, Schrader et al. 2013). Such a filter routine has been proposed e.g., by Peters et al. (2014).

6.6.5. Related methods

- Field: suction cups, wick samplers, pan lysimeters, lysimeters, raisin boxes, matric potential sensors, matrix potential sensors
- Laboratory: suction cups, wick samplers, lysimeters, matric potential sensors, matrix potential sensors

6.7. Substance-specific requirements of solute sampling

The different physico-chemical properties of the test substances in the soil solution cause a large spectrum of potential interactions between the materials used for sampling construction and the test substances regardless of the sampler type. In general, no sampling system enables the optimal sampling of all test substances at the same time, and therefore, the sampling system should be optimized to match the respective target compounds (Table 6.1). In most cases, the sampler material equilibrates with the soil solution with time, regardless of the chosen material. Nevertheless, we advise not to rely on equilibration in the field or in laboratory to achieve representative solute sampling, since i) the required time for equilibration is not well defined, and ii) the required time for equilibration cannot always be abide.

In long-term experiments, this equilibrium might indeed show that equilibrated suction cups do not affect the solution concentration of some solutes. But this is only valid when the concentration of this solute in the soil solution remains constant (Guggenberger and Zech 1992). However, as soon as the solute concentration in soils changes – and to determine precisely this, long-term experiments are required – a new equilibrium will be achieved (law of mass action). This balance adjustment shows that i) with an increasing solute concentration in soil, more ions are adsorbed at the filter surface or ii) with a decreasing solute concentration in soil, already adsorbed substances are released from the filter surface and pass into the percolate. In both cases, the result is a largely preventable variation of the soil water sample and the actual concentration course is either not noted, or at best with a delay.

In the following, recommendations for the most frequently sampled target compounds are provided. An overview is also presented in Table 6.1 without having the intention to be complete.

6.7.1. Protons (pH)

The partial pressure of CO_2 in the soil air has a decisive influence on the pH-value of the soil solution via the equilibrium of H_2CO_3, HCO_3^-, CO_3^{2-}, and carbon dioxide. Since the CO_2 partial pressure of the soil air is commonly several times greater than in the atmosphere, CO_2 will be released from soil solution samples in contact with the atmosphere (Zabowski and Sletten 1991). This increases the pH-value of the sampled soil solution by about 0.3 to 0.5 pH units (Suarez 1986 and 1987, Kaupenjohann and David 1996). To avoid outgassing of CO_2 Suarez (1986) proposed to use dual chamber sampling systems.

Furthermore, the pH-value of the sampled soil solution can be reduced by the oxidation of Fe^{2+}, Mn^{2+}, or NH_4^+ in the sampling system.

6.7.2. Nutrients (N, P, K, S, Ca, Mg)

The anion nitrate (NO_3^-) is characterized by low interaction with most sampler materials. Studies indicate good agreement of nitrate fluxes determined with lysimeters and suction cups for well-drained loamy sand and sandy loam (Goulding and Webster 1992, Webster et al. 1993). In contrast, ammonium (NH4+) is subject to cation exchange processes. Ammonium and nitrate concentrations are easily altered by biological processes (e.g., nitrification, N-assimilation). To avoid biological alterations short sampling intervals (<2 weeks), cool and dark storage of the samples in the field are therefore decisive. The sample solution can be additionally stabilized by adding acids (e.g., HCl) into the sampling bottles if this does not disturb the detection of other target data (e.g., pH-value, Cl$^-$) and their subsequent analysis. During transport samples should be kept cool and dark.

The phosphate ion has a strong affinity to metal hydroxides to which it is specifically bound by ligand exchange. Ceramic materials containing aluminum (hydr)oxide sorb and desorb considerable quantities of PO_4^{3-} (Bottcher et al. 1984). Sorption and desorption of PO_4^{3-} in suction cups and plates can be prevented by using porous PTFE (polytetrafluoroethylene, e.g., Teflon®) or glass, whereby porous PTFE often exhibit low air entrance values due to its hydrophobic surface (Bottcher et al. 1984). In general, a significant fraction of the phosphate is translocated by preferential flow events (Anderson and Xia 2001, Heckrath et al. 1995, Julich et al. 2017), and therefore, sampling systems with large surface areas such as lysimeters, pan lysimeters, capillary wick samplers, or suction plates, seem to be more suitable compared to suction cups.

The concentrations of K^+, Ca^{2+}, and Mg^{2+} are influenced by cation exchange. Due to the low cation exchange capacity of stainless steel and glass sinter materials, equilibrium between these cations and the materials soon establishes. On the other hand, carbon dioxide outgassing may result in the precipitation of calcium carbonate ($CaCO_3$) or magnesium carbonate ($MgCO_3$) in the sampling systems (Schwartz and Mielich 1993). New ceramic and glass components may also contain considerable quantities of Ca^{2+}, Mg^{2+}, K^+, and Na^+, and should therefore be cleaned by rinsing with hydrochloric acid (0.1 N) and deionized water (Grover and Lamborn 1970, Wessel-Bothe et al. 2000).

6.7.3. Trace metals (Al, Mn, Fe, Pb, Cr, Cu, Ni, Zn)

Aluminum concentration and aluminum speciation are controlled by the ambient soil pH. Changes in the soil pH-value induced by soil water extraction should therefore be considered and minimized. Aluminum often forms complexes with organic compounds. Therefore, the sampling systems should be characterized by low sorption of dissolved organic substances. Most ceramic materials contain Al (hydr)oxides and are therefore only suitable to a limited extent. Nylon, PE, and PTFE are more suitable for aluminum sampling).

Dissolved iron (Fe^{2+}) and manganese (Mn^{2+}) occur in environments with low redox potentials and are oxidized to Fe- and Mn-oxides after contact with oxygen in the sampling system (Schwartz and Miehlich 1993). Increase in pH also leads to the precipitation of Fe- and Mn-(hydr)oxides. In order to minimize the oxidation and precipitation of Fe^{2+} and Mn^{2+}, it is recommended that gas-tight tubing (e.g., of nylon) should be used, and the soil solution in suction cups should be sampled in the shaft. Another possible method is to flood the internal sampling system with inert gasses as N_2 prior to the suction phase. An acidification of the collected soil solution by adding acids (e.g., HCl, HNO_3) to pH 2 also stabilizes the Fe^{2+} concentrations.

Interactions of heavy metals with sampler materials were summarized by Wenzel and Wieshammer (1995). Heavy metals are specifically sorbed to metal hydroxides. Porous oxide ceramics are therefore only suitable for determining concentrations of heavy metals after long equilibration. This state may not be reached if low concentrations are to be detected. Especially, Cu and Pb were sorbed by ceramic materials in tests of Wenzel et al. (1997). Consequently, often membranes or porous bodies of organic polymers are recommended for sampling of trace metals. Not all samplers made from organic polymeric materials are equally suited for collecting trace metals. Rais et al. (2006) reported that suction cups made from a mixture of PTFE and a silicate effectively adsorbed Cu and Pb. Copper was also adsorbed by a pure organic polymer used for the production of micro suction cups (Rais et al. 2006). Stainless steel contains a large number of heavy metals (e.g., chromium, vanadium) and should not be used for this reason. More suitable materials are PE, PTFE, and nylon (Wessel-Bothe et al. 2000). The speciation of heavy metals depends on the pH-value. Changes in the pH-value due to outgassing of CO_2 therefore can influence the measured heavy metal concentrations. Furthermore, in the course of a precipitation of carbonates or oxides a co-precipitation of heavy metals in the measuring system may occur (Schwartz and Miehlich 1993). Heavy metals are complexed by dissolved organic compounds, which increase their overall concentration in the soil solution. Suitable sampling systems should therefore not sorb or release dissolved organic matter in large quantities. In tests performed by Rais et al. (2006) the presence of dissolved organic matter commonly reduced the adsorption of trace metals by various suction cup materials.

6.7.4. Organic compounds (pesticides, organic pollutants, xenobiotics, dissolved organic carbon, organically bound nutrients)

Pesticides, xenobiotics, and organic pollutants such as polycyclic aromatic hydrocarbons (PAHs) and polychlorinated biphenyls (PCBs), display a wide range of physico-chemical properties (e.g., water solubility and vapor pressure). It is therefore difficult to make universal and comprehensive recommendations for this group of substances. Depending on their water solubility, the majority of substances are in partitioning equilibrium between organic and aqueous phase. Plastics, adhesives, and elastomers absorb organic compounds and are not well suited for the construction of sampling systems for these substances. Glass (Wessel-Bothe et al. 2000) and stainless steel are more suitable.

Depending on the vapor pressure of the compound, volatilization from the sample of soil solution may occur. In this case, dual-chamber systems should be used (Suarez 1986 and 1987, Wood et al. 1981).

Dissolved organic carbon (DOC) or more precisely dissolved organic matter (DOM) is specifically adsorbed to metal hydroxides via its carboxyl and hydroxyl groups. Many ceramic materials therefore sorb considerable quantities of DOM and are only suitable for determining DOC concentrations after a long period of equilibration (Guggenberger and Zech 1992, Wessel-Bothe et al. 2000). Porous glass (Wessel-Bothe et al. 2000) or stainless steel is more suitable. Adhesives, glues, and elastomers used for the construction of suction cups and plates contain solvents and plasticizers, which are released into the sampled soil solution and may increase the measured DOC concentration (Siemens and Kaupenjohann 2003). The sampled solution should therefore be protected from contact with adhesives or elastomers. Stainless-steel capillaries and PTFE tubes are thus more appropriate as sampling device material. Since organically bound nutrients are naturally subjected to the same processes as DOM, it is also recommended that glass, stainless steel, and PTFE should be used.

Preferential flow represents the most significant transport mechanism for strongly sorbing pesticides, organic pollutants, DOM, and organically bound nutrients (Jene 1998, Kaiser et al. 2000, Qualls 2000, Vereecken 2005). For this reason, large sampling systems such as lysimeters, wick samplers, or suction plates are appropriate for measuring representative concentrations and fluxes.

6.7.5. Colloids and microorganisms

A major problem of sampling colloids and microorganisms is the size exclusion by filtration (Bell 1974, Dazzo and Rothwell 1974). In most cases, colloids, bacteria, and viruses are therefore sampled with zero-tension systems (Kaplan et al. 1993 and 1996, Thompson and Scharf 1994) or systems with coarse pores (pore diameter >16 μm) (Schäfer et al. 1998). Cigány et al. (2005) tested the suitability of fiberglass wicks for sampling various kinds of specific colloids (feldspathoids, ferrihydrite, montmorrillonite, and kaolinite) and of a mixture of colloids extracted from coarse sand at various flow rates and pH under unsaturated conditions. The permeability of the wicks for colloids depended strongly on pH, flow rate, and type of colloid. The authors conclude that colloids can be retained in the wicks under many conditions, so that their use for colloid sampling should be considered with caution. Similarly, Shira et al. (2006) reported a considerable retention of negatively charged silica microspheres and ferrihydrite in fiberglass wicks. They concluded that wicks are probably suited to characterize mobile colloids in the vadose zone qualitatively but could not be expected to sample colloid concentrations quantitatively. Ilg et al. (2007) compared the efficiencies of a zero-tension system, membranes of various mesh size, glass fiber wicks, and porous glass plates for sampling radioactively labelled goethite in a column experiment. They concluded that nylon membranes of 16 μm mesh and zero tension are best suited for sampling goethite colloids. Overall, there is still limited experience with respect to sampling of colloids under unsaturated conditions. Retention of colloids in sampling systems strongly depends on the type of colloids, geochemical, and on hydrodynamic conditions that prevail. Hence, individual testing of colloid retention in the sampling system to be used seems inevitable.

6.8. References

Abdou, H.M., Flury, M., 2004. Simulation of water flow and solute transport in free- drainage lysimeters and field soils with heterogeneous structure. Eur. J. Soil Sci. 55: 229–241.

Anderson, R., Xia, L., 2001. Agronomic measures of P, Q/I parameters and lysimeter collectable P in subsurface soil horizons of a long-term slurry experiment. Chemosphere 42: 171–178.

Bell, R., 1974. Porous ceramic soil moisture samplers, an application in lysimeters studies on effluent spray irrigation. N. Z. J. Exp. Agric. 2: 173–175.

Ben-Gal, A., Shani, U., 2002. A highly conductive drainage extension to control the lower boundary condition of lysimeters. Plant Soil 239: 9–17.

Bergström, L.F., 1990. Use of lysimeters to estimate leaching of pesticides in agricultural soils. Environ. Pollut. 67: 325–347.

Bethge-Steffens, D, Meissner, R., Rupp, H., 2004. Development and practical test of a weighable groundwater lysimeter for floodplain sites. J. Plant Nutr. Soil Sci. 167: 516–524.

Bloem, E., Hagervorst, F.A.N., de Rooij, G.H., 2009. A field experiment with variable-suction multi-compartment samplers to measure the spatio-temporal distribution of solute leaching in an agricultural soil. Journal of Contaminant Hydrology 105: 131–145.

Bloem, E., Hogervorst, F.A.N., de Rooij, G.H., Stagnitti, F., 2010. Variable-suction multicompartment samplers to measure spatiotemporal unsaturated water and solute fluxes. Vadose Zone Journal 9 (1): 148–159.

Boll, J., Steenhuis, T.S., Selker, J.S., 1992. Fiberglass wicks for sampling of pore water and solutes in the vadose zone. Soil Sci. Soc. Am. J. 56: 701–707.

Bottcher, A.B., Miller, L.W., Campbell, K.L., 1984. Phosphorous adsorption on various soil-water extraction cup material. Effect of acid wash. Soil Sci. 137: 239–245.

Brandi-Dohrn, F.M., Dick, R.P., Hess, M., Selker, J.S., 1996. Field evaluation of passive capillary samplers. Soil Sci. Soc. Am. J. 60: 1705–1713.

Briggs, L., McCall, A., 1904. An artifical root for including capillary movement of soil moisture. Science 20: 566–569.

Chiu, T.F., Shackelford, C.D., 2000. Laboratory evaluation of sand under-drains. J. Geotech. Geoenviron. Engin. 126: 990–1001.

Cigány, S., Flury, M., Harsh, J.B., Williams, B.C., Shira, J.C., 2005. Suitability of fiberglass wicks to sample colloids from vadose zone pore water. Vadose Zone J. 4: 175–183.

Ciglasch, H., Amelung, W., Totrakool, S., Kaupenjohann, M., 2005. Water flow patterns and pesticide fluxes in an upland soil in northern Thailand. Eur. J. Soil Sci. 56: 765–777.

Colman, E.A., 1946. A laboratory study of lysimeter drainage under controlled soil moisture tension. Soil Sci. 62: 365–382.

Dazzo, F.B., Rothwell, D.F., 1974. Evaluation of porcelain cup soil water samplers for bacteriological sampling. Applied Microbiol. 27: 1172–1174.

Dowdell, R.J., Webster, C.P., 1980. A lysimeter study using nitrogen N-15 on the uptake of fertilizer nitrogen by perennial ryegrass swards and losses by leaching. J. Soil Sci. 31: 65–75.

Duke, H., Haise, H., 1973. Vacuum extractors to assess deep percolation losses and chemical constituents of soil water. Soil Sci. Soc. Am. Proc. 37: 963–964.

Ebermayer, E., 1873. Die physikalische Bedeutung des Waldes auf Luft und Boden und seine klimatologische und hygienische Bedeutung. Wiegandt, Hempel and Parey, Berlin.

Fank, J., 2013. Wasserbilanzauswertung aus Präzisionslysimeterdaten, in: 15. Gumpensteiner Lysimetertagung 2013, Lehr- und Forschungszentrum für Landwirtschaft Raumberg-Gumpenstein, Irdning, Austria, 85–92.

Flemming, J., Butters, G., 1995. Bromide transport detection in tilled and nontilled soil: Solution sampler vs. soil cores. Soil Sci. Soc. Am. J. 59: 1207–1216.

Flury, M., Yates, M.V., Jury, W.A., 1999. Numerical analysis of the effect of the lower boundary condition on solute transport in lysimeters. Soil Sci. Soc. Am. J. 63: 1493–1499.

Führ, F., Burauel, P., Dust, M., Mittelstaedt, W., Pütz, T., Reinken, G., Stork, A., 1998. Comprehensive tracer studies on the environmental behaviour of pesticides: the lysimeter concept. In: Führ, F., Hance, R.J., Plimmer, J.R., Nelson, J.O. (eds.), The Lysimeter Concept – Environmental Behavior of Pesticides. Am. Chem. Soc. Washington DC, ACS Symp. Ser. 699: 1–20.

Gee, G.W., Newman, B.D., Green, S.R., Meissner, R., Rupp, H., Zhang, Z.F., Keller, J.M., Waugh, W.J., van der Velde, M., Salazar, J., 2009. Passive wick fluxmeter: Design considerations and field applications. Water Resources Research 45. doi: 10.1029/2008WR007088

Gee, G.W., Ward, A.L., Caldwell, T.G., Ritter, J.C., 2002. A vadose-zone water fluxmeter with divergence control. Water Resour. Res. 38: doi: 10.1029/2001WR00816.

Gee, G.W., Zhang, Z.F., Ward, A.L., 2003. A modified vadose-zone fluxmeter with solution collection capability. Vadose Zone J. 2: 627–632.

Giessler, R., Lundström, U.S., Grip, H., 1996. Comparison of soil solution chemistry assessment using zero-tension lysimeters or centrifugation. Eur. J. Soil Sci. 47: 395–405.

Goulding, K.W., Webster, C.P., 1992. Methods for measuring nitrate leaching. Aspects Applied Biol. 30: 63–70.

Goyne, K.W., Day, R.L., Chorover, J. 2000. Artifacts caused by collection of soil solution with passive capillary samplers. Soil Sci. Soc. Am. J. 64: 1330–1336.

Grossmann, J., Bredemeier, M.,. Udluft, P., 1990. Sorption of trace elements by suction cups of aluminum-oxide, ceramic and plastics. Zeitschrift für Pflanzenernähr. Bodenkd. 153: 359–364.

Grossmann, J., Freitag, G., Merkel, B., 1985. Eignung von Nylon- und Polyvinylfluorid-Membranfiltern als Material zum Bau von Saugkerzen. Z. Wasser- und Abwasserf. 18: 187–190.

Grossmann, J., Merkel, B., Udluft, P., 1988. Carbonate equilibrium in soil water samples. Z. Wasser- und Abwasserf. 21: 177–181.

Grossmann, J., Udluft, P., 1991. The extraction of soil water by the suction-cup method: A review. J. Soil Sci. 42: 83–93.

Grover, B.L., Lamborn, R.E. 1970. Preparation of porous ceramic cups to be used for extraction of soil water having low solute concentrations. Soil Sci. Soc. Am. Proc. 34: 706–708.

Guggenberger, G., Zech, W., 1992. Sorption of dissolved organic carbon by ceramic P80 suction cups. Zeitschrift für Pflanzenernähr. Bodenkd. 155: 151–155.

Guggenberger, G., Zech, W., 1993. Zur Dynamik gelöster organischer Substanzen (DOM) in Fichtenökosystemen – Ergebnisse analytischer DOM-Fraktionierung. Zeitschrift für Pflanzenernähr. Bodenkd. 156: 341–347.

Hädrich, F., Stahr, K., Zöttl, H.W., 1977. Die Eignung von Al2O3-Keramikplatten und Ni-Sinterkerzen zur Gewinnung von Bodenlösungen für die Spurenelementanalyse. Mitteilung Dtsch. Bodenkd. Gesellschaft. 25: 151–162.

Hansen, E.A., Harris, A.R., 1975. Validity of soil-water samples collected with porous ceramic cups. Soil Sci. Soc. Am. Proc. 39: 528–536.

Hart, G.L., Lowery, B., 1997. Axial-radial influence of porous cup soil solution samplers in a sandy soil. Soil Sci. Soc. Am. J. 61: 1765–1773.

Heckrath, G., Brookes, P.C., Poulton, P.R., Goulding, K.W.T., 1995. Phosphorus leaching from soils containing different phosphorus concentrations in the Broadbalk experiment. J. Environ. Qual. 24: 904–910.

Holder, M., Brown, K.W., Thomas, J.C., Zabcik, D., Murray, H.E., 1991. Capillary wick unsaturated pore water sampler. Soil Sci. Soc. Am. J. 55: 1195–1202.

Ilg, K., Ferber, E., Stoffregen, H., Winkler, A., Pekdeger, A., Kaupenjohann, M., Siemens, J., 2007. Unsaturated colloid transport through columns with differing sampling systems Soil Sci. Soc. Am. J. 71: 298–305.

Jemison, M.J., Fox, R.H., 1992. Estimation of zero-tension pan lysimeter collection efficiency. Soil Sci. 154: 85–94.

Jene, B. 1998. Transport of bromide and benazolin in lysimeters and a field plot with grid suction bases in a sandy soil. Doctoral Dissertation, University of Hohenheim, Germany.

Julich, D., Julich, S., Feger, K.-H., 2017. Phosphorus fractions in preferential flow pathways and soil matrix in hillslope soils in the Thuringian Forest (Central Germany). J. Plant Nutr. Soil Sci. 2017, 180: 407–417.

Jury, W., Flühler, H., 1992. Transport of chemicals through soils: Mechanisms, models, and field applications. Adv. Agron. 47: 141–201.

Kaiser, K., Guggenberger, G., Zech W., 2000. Organically bound nutrients in dissolved organic matter fractions in seepage and pore water of weakly developed forest soils. Acta Hydrochim. Hydrobiol. 28: 411–419.

Kaplan, D.I., Bertsch, P.M., Adriano, D.M., Miller W.P., 1993. Soil-borne mobile colloids as influenced by water flow and organic carbon. Environ. Sci. Technol. 27: 1193–1200.

Kaplan, D.I., Sumner, M.E., Bertsch, P.M., Adriano, C., 1996. Chemical conditions conducive to the release of mobile colloids from ultisol profiles. Soil Sci. Soc. Am. J. 60: 269–274.

Kasteel, R., Pütz, T., Vanderborght, J., Vereecken, H., 2010. Fate of Two Herbicides in Zero-Tension Lysimeters and in Field Soil. J. Environ. Qual. 39: 1451–1466.

Kasteel, R., Pütz, T., Vereecken, H. 2007. An experimental and numerical study on flow and transport in a field soil using zero-tension lysimeters and suction plates. European Journal of Soil Science 58: 632–645.

Kaupenjohann, M., David, M.B., 1996. Evidence for effects of CO_2 on soil solution chemistry in a Spodosol by a simple in-field extractor. Zeitschrift für Pflanzenernähr. Bodenkd. 159: 195–198.

Knutson, J.H., Selker, J.S., 1994. Unsaturated hydraulic conductivities of fiberglass wicks and designing capillary wick pore water samplers. Soil Sci. Soc. Am. J. 58: 721–729.

Kosugi, K., Katsuyama, M., 2004. Controlled-suction period lysimeter for measuring vertical water flux and convective chemical fluxes. Soil Sci. Soc. Am. J. 68: 371–382.

Krone, R., Ludwig, H., Thomas, J., 1951. Porous tube device for sampling soil solution during water spreading operations. Soil Sci. 73: 211–219.

Leinemann, T., Mikutta, R., Kalbitz, K., Schaarschmidt, F., Guggenberger, G., 2016. Small scale variability of vertical water and dissolved organic matter fluxes in sandy Cambisol subsoils as revealed by segmented suction plates. Biogeochemistry 131: 1–15.

Linden, D., 1977. Design, installation, and use of porous ceramic samplers for monitoring soil water quality. Agric. Res. Stn., U.S. Dep. Agric., Tech. Bull. 1562.

Louie, M.J., Shelby, P.M., Smesrud, J.S., Gatchell, L.O., Selker, J.S. 2000. Field evaluation of passive capillary samplers for estimating groundwater recharge. Water Resour. Res. 36: 2407–2416.

Magid, J., Christensen, N., Nielsen, H., 1992. Measuring phosphorous fluxes through the root zone of a layered sandy soil: Comparison between lysimeter and suction cell solution. J. Soil Sci. 43: 739–747.

Marques, R., Ranger, J., Gelhaye, D., Pollier, B., Ponette, Q., Goedert O., 1996. Comparison of chemical composition of soil solutions collected by zero-tension plate lysimeters with those from ceramic-cup lysimeters in a forest soil. Eur. J. Soil Sci. 47: 407–417.

McGuire, P.E., Lowery, B., Helmke, P., 1992. Potential sampling error: trace metal adsorption on vacuum porous cup samplers. Soil Sci. Soc. Am. J. 56: 74–82.

McGuire, P.E., Lowery, B.,1994. Monitoring drainage solution concentrations and solute flux in unsaturated soil with a porous cup sampler and soil moisture sensors. Ground Water. 32: 356–362.

Meissner, R., Rupp, H., Schubert, M., 2000. Novel lysimeter techniques – a basis for the improved investigation of water, gas, and solute transport in soils. J. Plant Nutr. Soil Sci. 163: 603–608.

Meissner, R., Seeger, J., Rupp, H., Seyfarth, M., Borg, H., 2007. Measurement of dew, fog, and rime with a high precision gravitation lysimeter. Journal of Plant Nutrition and Soil Science 170: 335–344.

Mertens, J., Barkle, G.F., Stenger, R.. 2005. Numerical analyses to investigate the effect of the design and installation of equilibrium tension plate lysimeters on leachate. Vadose Zone J. 4: 488–499.

Mertens, J., Diels, J., Feyen, J., Vanderborght, J., 2007. Numerical analysis of passive capillary wick samplers prior to field installation. Soil Sci. Soc. Am. J. 71: 35–42.

Mertens, J., Tuts, V., Diels, J., Vanderborght, J., Feyen, J., Merckx, R., 2008. Design and Testing of a Drop Counter for Use in Vadose Zone Water Samplers. Vadose Zone J. 7: 434–438.

Mitchell, M.J., McGee, G., McHale, P., Weathers, K., 2001. Experimental design and instrumentation for analyzing solute concentrations and fluxes for quantifying biogeochemical processes in watersheds. p. 15–21. In: Methodology Paper Ser. of the 4th Int. Conf. on ILTER in East Asia and Pacific Region, Ulaanbaatar-Hatgal, Mongolia.

Nehls, T., Nam Rim, Y., Wessolek, G., 2011. Technical note on measuring run-off dynamics from pavements using a new device: the weighable tipping bucket. Hydrology and Earth System Sciences. 15: 1379–1386.

Parizek, R., Lane, B., 1970. Soil-water sampling using pan and deep pressure-vacuum lysimeters. J. of Hydrol. 11: 1–21.

Peters, A., Nehls, T., Schonsky, H, Wessolek, G., 2014. Separating precipitation and evapotranspiration from noise – a new filter routine for high-resolution lysimeter data. Hydrology and earth System Sciences 18: 1189–1198.

Qualls, R.G., 2000. Comparison of the behaviour of soluble organic and inorganic nutrients in forest soils.

For. Ecol. Manage. 138: 29–50.
Rais, D., Nowack, B., Schulin, R., Luster, J., 2006. Sorpton of trace metals by standard and micro suction cups in the absence and presence of dissolved organic carbon. J. Environ. Qual.35: 50–60.
Ramier, D., Berthier, E., Andrieu, H., 2004. An urban lysimeter to assess runoff losses on asphalt concrete plates. Phys. Chem Earth. 29: 839–847.
Ranger, J., Marques, R., Jussy, J.-H., 2001. Forest soil dynamics during stand development assessed by lysimeter and centrifuge solutions. For. Ecol. Manage. 144:129–145.
Rhoades, J.D., Oster, J.D., 1986. Solute Content. p. 985–1006. In: Klute, A. (ed.) Methods of Soil Analysis, Part 1, Second edition, SSSA Book Series, 5, SSSA, Madison, WI.
Richards, L.A., 1950. Laws of soil moisture. Trans. Am. Geophys. Union. 31: 750–756.
Rimmer, A., Steenhuis, T.S., Selker, J.S., 1995. One-dimensional model to evaluate the performance of wick samplers in soil. Soil Sci. Soc. Am. J. 59: 88–92.
Schäfer, A., Ustohal, P., Harms, H. Stauffer, F., Dracos, T., Zehnder, A.J.B., 1998. Transport of bacteria in unsaturated porous media. J. Contam. Hydrol. 33: 149–169.
Schoen, R., Gaudet, J.P., Bariac, T., 1999. Preferential flow and solute transport in a large lysimeter study under controlled boundary conditions. Journal of Hydrology 215: 70–81.
Schrader, F., Durner, W., Fank, J., Gebler, S., Pütz, T., Hannes, M., Wollschläger, U., 2013. Estimating precipitation and actual evapotranspiration from precision lysimeter measurements. In: Romano, N., D'Urso, G., Severino, G., Chirico, G., Palladino, M. (Eds.), Four Decades of Progress in Monitoring and Modeling of Processes in the Soil-Plant-Atmosphere System: Applications and Challenges. Procedia Environmental Sciences: 543–552
Schwarz, R., Miehlich, G., 1993. Einsatz der Saugkerzentechnik in reduzierten Horizonten: Stoffausfällung im Fördersystem und methodische Verbesserung. Mitteilung Dtsch. Bodenkd. Gesellschaft. 72: 457–460.
Severson, R., Grigal, D., 1976. Soil solution concentrations: Effects of extraction time using porous ceramic cups under constant tension. Water Resour. Bull. 12: 1161–1170.
Shira, J.M., Williams, B.C., Flury, M., Czigány, S., Tuller, M., 2006. Sampling silica and ferrihydrite colloids with fiberglass wicks under unsaturated conditions. J. Environ. Qual. 35: 1127–1134.
Siemens, J., Kaupenjohann, M., 2004. Comparison of three methods for field measurement of solute leaching in a sandy soil. Soil Sci. Soc. Am. J. 68: 1191–1196.
Suarez, D.L., 1986. A soil water extractor that minimizes CO_2 degassing and pH errors. Water Resourc. Res. 22: 876–880.
Suarez, D.L., 1987. Prediction of pH errors in soil water extractors due to degassing. Soil Sci. Soc. Am. J. 51: 64–67.
Thompson, M.L., Scharf, R.L., 1994. An improved zero-tension lysimeter to monitor colloid transport in soils. J. Environ. Qual. 23: 378–383.
Unold, G. von Fank, J., 2008. Modular design of field lysimeters for specific application needs. Water Air and Soil Pollution Focus. 8: 233–242.
van der Velde M, Gee, G.W., Green, S., 2003. After fire comes rain: Drainage in Tonga. WISPAS 86: 1–2.
van der Velde, M., Gee, G.W., Green, S., 2003. After fire comes rain: Drainage in Tonga. WISPAS 86: 1–2.
van der Velde, M., Green, S.R., Gee, G.W., Vanclooster, M., Clothier, B.E., 2004. Evaluation of drainage from passive suction and nonsuction flux meters in a volcanic clay soil under tropical conditions. Vadose Zone J. 4: 1201–1209.
van Grinsven, J.J.M., Boolting, H.W.G., Dirksen, C., van Breemen, N., Bongers, N., Waringa, N., 1988. Automated in situ measurement of unsaturated soil water flux. Soil Sci. Soc. Am. J. 52: 1215–1218.
van Weesenbeck, I., Schabacker, D.J., Winton, K., Heim, L., Winberry, M.W., Williams, M.D.J., Weber, B., Swain, L.R., Velagaleti, R., 1998. Demonstration of the functionality of a self-contained modular lysimeter design for studying the fate and transport of chemicals in soil under field conditions. In: Führ, F., Hance, R.J., Plimmer, J.R., Nelson, J.O. (eds.), The Lysimeter Concept – Environmental Behavior of Pesticides. Am. Chem. Soc. Washington DC, ACS Symp. Ser. 699: 122–135.
Vereecken, H., 2005. Mobility and leaching of glyphosate: A review. Pest Managm. Sci. 61: 1139–1151
Vereecken, H., Dust, M., 1998. Modelling water flow and pesticide transport at lysimeter and field scale. ACS Symposium Series 699: 189–202.
Wagner, G., 1962. Use of porous ceramic cups to sample soil water within the profile. Soil Sci. 94: 379–386.
Warrick, A., Amoozegar-Fard, A., 1977. Soil water regimes near porous water samplers. Water Resourc.

Res. 13: 203–207.
Webster, C.P., Shepherd, M.A., Goulding, K.W.T., Lord, E.I., 1993. Comparison of methods for measuring the leaching of mineral nitrogen from arable land. J. Soil Sci. 44: 49–62.
Weihermüller, L., Kasteel, R., Vanderborght, J., Pütz, T., Vereecken, H., 2005. Spatial impact of soil water extraction with a suction cup – Results of numerical Simulations. Vadose Zone J. 4: 899–907.
Weihermüller, L., Kasteel, R., Vanderborght, J., Šimůnek, J., Vereecken, H., 2011. Uncertainties in pesticide monitoring using suction cups: Evidences from numerical simulations. Vadose Zone Journal 10: 1287–1298.
Weihermüller, L., Kasteel, R., Vereecken, H., 2006. Effects of soil heterogeneity on solute breakthrough sampled with suction cups: Results of numerical simulations. Vadose Zone J. 5: 886–893.
Weihermüller, L., Siemens, J., Deurer, M., Knoblauch, S., Rupp, H., Göttlein, A., Pütz, T., 2007. In-situ soil water extraction: A Review. Journal of Environmental Quality 36: 1735–1748.
Weihermüller, L., Siemens, J., Deurer, M., Knoblauch, S., Rupp, H., Göttlein, A., Pütz, T., 2007. In-situ soil water extraction: A Review. Journal of Environmental Quality 36: 1735–1748
Wenzel, W., Sletten, R.S., Brandstetter, A., Wieshammer, G., Stingeder, G., 1997. Adsorption of trace metals by teinsion lysimeters: Nylon membrane vs. ceramic cup. J. Environ. Qual. 26: 1430–1434.
Wenzel, W., Wieshammer, G., 1995. Suction cup materials and their potential to bias trace metal analyses of soil solutions: A review. Int. J. Environ. Anal. Chem. 59: 277–290.
Wessel-Bothe, S., Pätzold, S., Klein, C., Behre, G., Welp, G., 2000. Adsorption von Pflanzenschutzmitteln und DOC an Saugkerzen aus Glas und Keramik. J. Plant Nutr. Soil Sci. 163: 53–56.
Wolters, A., Leistra, M., Linnemann, V., Smelt, J.H., van den Berg, F. Klein, M., Jarvis, N., Boesten, J.J.T.I., Vereecken, H., 2003. Pesticide volatilisation from plants: Improvement of the PEARL, PELMO, and MACRO models. P. 985–994. In: Del Re, A.A.M., Capri, E., Padovani, L., Trevisan, M. (eds.), Pesticide in Air, Plant, Soil and Water System, Proc. XII Symp. Pestic. Chem. June 4–6, Piacenza, Italy.
Wood, A.L., Wilson, J.T., Cosby, R.L., Hornsby, A.G., Baskin, L.B., 1981. Apparatus and procedure for sampling soil profiles for volatile organic compounds. Soil Sci. Soc. Am. J. 45: 442–444.
Yang, J.F., Li, B.Q., Liu, S.P. 2000. A large weighting lysimeter for evaporation and soil water-groundwater exchange studies. Hydrological Processes. 14: 1887–1897.
Zabowski, D., Sletten, R.S., 1991. Carbon dioxide degassing effects on the pH of Spodosol soil solutions. Soil Sc. Soc. Am. J. 55: 1456–1461.
Zhu, Y, Fox, R.H., Toth, J.D., 2002. Leachate collection efficiency of zero-tension pan and passive capillary fiber wick lysimeters. Soil Sci. Soc. Am. J. 66: 37–43.

7. Infiltration and water conductivity in saturated and unsaturated soils

R. Schneider, J. Rodrigo-Comino, D. Demand, H. P. Schrey

7.1. Introduction

The infiltration rate of water into soils and their saturated and unsaturated hydraulic conductivity are of great relevance, e.g., for a soil's ability to supply water to plants, for processes that redistribute both nutrients and pollutants within soils, and for assessing relevant runoff processes and soil water movement (surface runoff, subsurface flow, deep percolation). Furthermore, they are central for the development of floods, for soil erosion processes, and groundwater recharge.

The infiltration capacity of soil surfaces may vary greatly, depending on current boundary conditions (e.g., soil type and structure, soil use, soil water content, and intensity of precipitation). Only part of the water infiltrated into a soil will regularly be stored there, while fractions of the infiltrated water move into lower sections of the soil profile at various rates.

The rate of infiltration of rain and surface water into a soil may only be quantified in rare cases by soil's Ks (saturated hydraulic conductivity) only. In most cases, soils are unsaturated at the onset of infiltration; a layer near the surface may, however, saturate quickly in the course of strong precipitation.

Even then, water flow in a saturated soil at the surface may not be characterized by its saturated hydraulic conductivity (Ks), because the front of water propagating the soil might enter unsaturated regions, which may feature greatly varying hydraulic conductivities.

This has the effect, that the infiltration front close to the soil surface, may proceed and features preferential flow (fingering or funneling), due to soil heterogeneity. Soil layers, close to water saturation may entrap soil air and impede pressure equilibration, which is necessary for further infiltration into the soil. Other interfering effects, such as hydrophobicity, may limit the ability to infer the actual infiltration rates from measured hydraulic conductivities.

The hydraulic conductivity is also of great importance for predicting the extent and the rate of distribution of pollutants both in soils and the groundwater. All such measurements must consider the scale at which they are conducted. In the laboratory, most measurements are conducted on relatively small soil sample volumes (commonly 100 – 250 cm^3) compared to many field measurement setups (volumes in the order of dm^3), and irrigation experiments (volumes in the order of m^3).

7.2. Method selection

The decision tree of Fig. 7.1 aims to help in selecting the general approach to measure infiltration and to estimate the (saturated) hydraulic conductivity. All methods shown are generally suited for measurements in flat to slightly inclined areas. Hints and limits to applying them in steeper field situations are presented in the sections discussing the particular method.

The first step is to decide if the measurement is to be conducted at the soil surface (infiltration) or within the soil profile (water or hydraulic conductivity). Infiltration measurements at the soil surface are fundamentally different in terms of the method of water supply. On the one hand, there are infiltration measurements (flooding at different water levels) e.g., single or double ring infiltrometers, and on the other hand, methods that use a water-tension controlled water supply (tension- or hood-infiltrometers).

Among water or hydraulic conductivity measurement methods in the soil, one can distinguish between methods applicable above the water table (e.g., Guelph-Permeameter) and others, which may be used on or close to the groundwater table (borehole-method). Additionally, we would like to point out, that one can differentiate between small-scale methods (shown in Fig. 7.1) and large-scale methods (e.g., irrigation experiments).

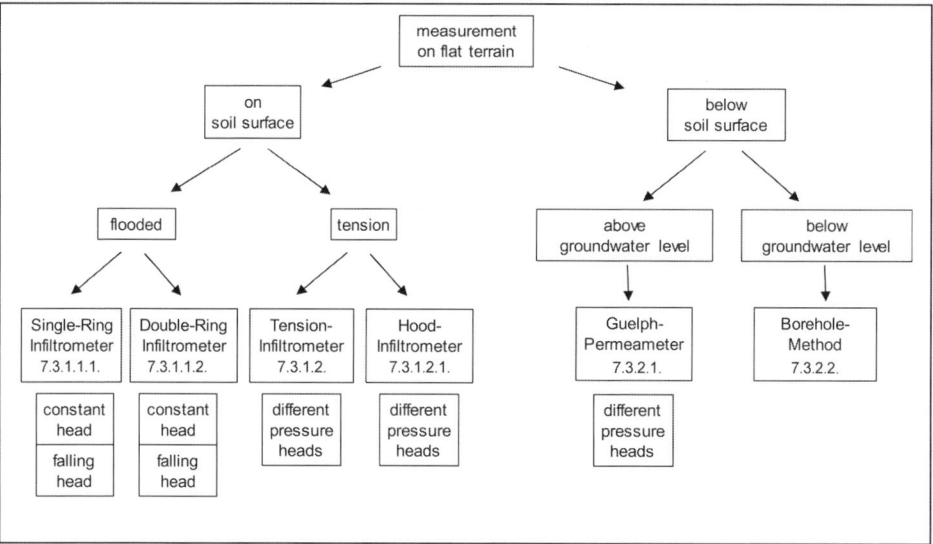

Fig. 7.1. Selection scheme for methods/techniques to measure the infiltration rate or the (saturated) hydraulic conductivity of soils. The numbers in the boxes denote the respective subsection that discuss the respective method in detail.

7.3. Method descriptions

7.3.1. Infiltration measurements

Commonly, infiltration measurements are conducted at flat to slightly inclined soil surfaces (Fig. 7.1, left). We distinguish different methods to infiltrate water; flooded infiltration or tension controlled infiltration.

7.3.1.1. Flooded Infiltration

According to DIN 19682 infiltration means "access of water through narrow channels in the lithosphere", a definition which also applies to the infiltration of water into the soil, the topmost layer of the earth, e.g., as the result of precipitation or flooding (ASTM D3385). A great number

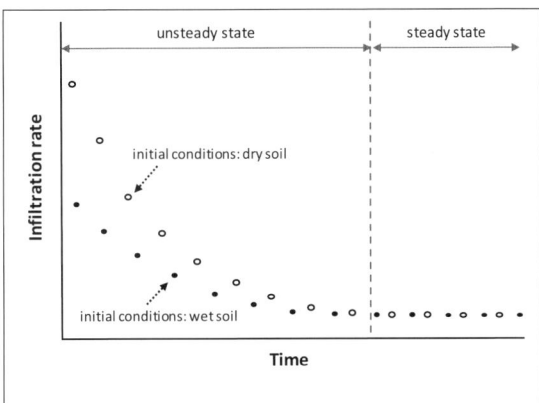

Fig. 7.2. Schematic diagram showing the change of the infiltration rate as a function of initial soil water content and time (modified from Durner 2011).

of variable factors, such as physical properties of the soil surface and within the soil (e.g., silting, porosity), the water content of the soil, the amount of water infiltrated per unit of time, vegetation cover, and anthropogenic factors (e.g., soil compaction) control the infiltration behavior of soils (Johnson 1963). Consequently, the infiltration behavior of soils is not constant, but varies greatly in space and time.

In general, the infiltration rate is defined as the amount of water a soil absorbs per unit of time.

The initial rate of infiltration, may, depending on the starting situation, be influenced by soil type and initial water content and may be orders of magnitude higher than its final, steady state, infiltration rate (Fig. 7.2). With increasing infiltration time (or water amount), the influence of relevant processes shifts from the variable capillary forces to gravitational forces, and coincides with a decrease of soil water potential gradient (Vereecken et al. 2019). The final infiltration rate is reached at a time, where the gravitational forces dominate the downward flow and matric potential gradients are negligable.

Finally, the cumulative infiltration is the total amount of water infiltrated into a soil within a defined period of time (Maniak 2010), which depend on the initial water content in the soil profile.

Under typically very variable field conditions, any single method is generally insufficient to properly characterize the infiltration behavior (Johnson 1963). To measure infiltration at the soil surface, most commonly, ponding ring infiltrometers (single and double ring types) or infiltrometers with water-tension controlled water supply (tension infiltrometers) are used (Mertens et al. 2002, Reynolds et al. 2000, Verbist et al. 2013).

In the following section the different methods will be introduced and discussed in detail.

7.3.1.1.1. Single-Ring Infiltrometer

The great advantage of single ring infiltrometry (Fig. 7.3), compared to other methods (e.g., double ring, hood, tension infiltrometer), is that it is simple to conduct with comparably little technical effort and above all quick (Di Prima 2016). A considerable disadvantage though, is the lateral subsurface flow of water below the ring, which is part of the infiltrated water volume and thus leads to overestimating a soils vertical (1-D) infiltration capacity. The mathematical derivation of the saturated hydraulic conductivity of soils thus requires considerably more complex analyzing routines (Torfs 2008, Verbist et al. 2013).

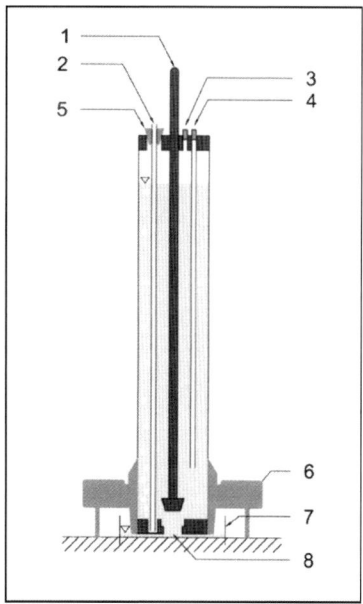

Fig. 7.3. Schematic diagram of an automatic single-ring infiltrometer after Di Prima et al. (2016). 1: piston, 2: air entry tube, 3: connector for vacuum side of pressure sensor, 4: connector for pressure side of pressure sensor, 5: rubber, 6: tripod, 7: water containing ring, 8: outlet.

Ring size

For measurements in the field, single ring infiltrometers with a large diameter should be used, because this reduces the bias of laterally flowing water at the ring margins and also integrates out small scale soil heterogeneities. According to Youngs (1987), measurements approach idealized conditions as ring radii of greater than 150 mm are used (this corresponds to radii approaching infinity) and average cumulative infiltration data converge (Fig. 7.4).

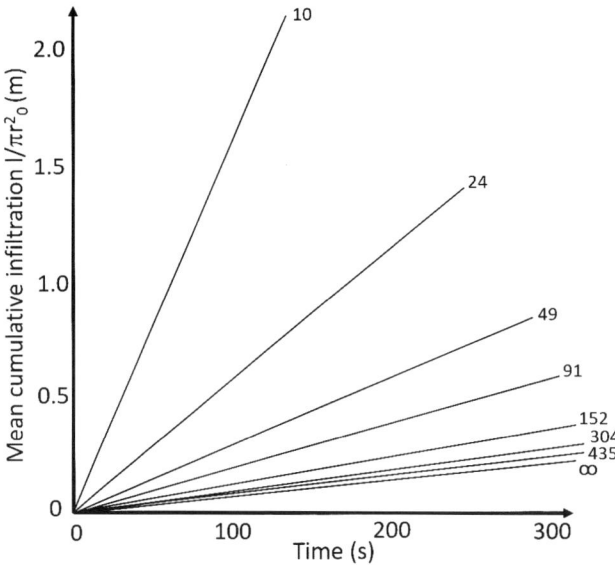

Fig. 7.4. Mean cumulative amounts of infiltration into a coarse sand as a function of the radius of the used infiltrometer ring (numbers on the lines in the diagram in mm). From Youngs (1987, p. 626).

The undesired effects of later flow and the overestimate of the integration capacity of a soil thus decrease with increasing diameter of the infiltrometer.

Water supply

Generally, we distinguish between two methods of supplying water during the measurement, the constant head method, (see Fig. 7.5A) and the falling head method (Fig. 7.5B). The constant head method keeps the water level over the entire measurement period, usually by using a Mariotte bottle (Di Prima 2016). This setup is more complex, but is advantageous for processing, modelling, and data interpreting.

The falling head method relies on measuring the water level (decrease) in the ring over time. The water in the ring can be either fully emptied or repeatedly refilled to the original level, while in both cases the water levels and times have to be recorded.

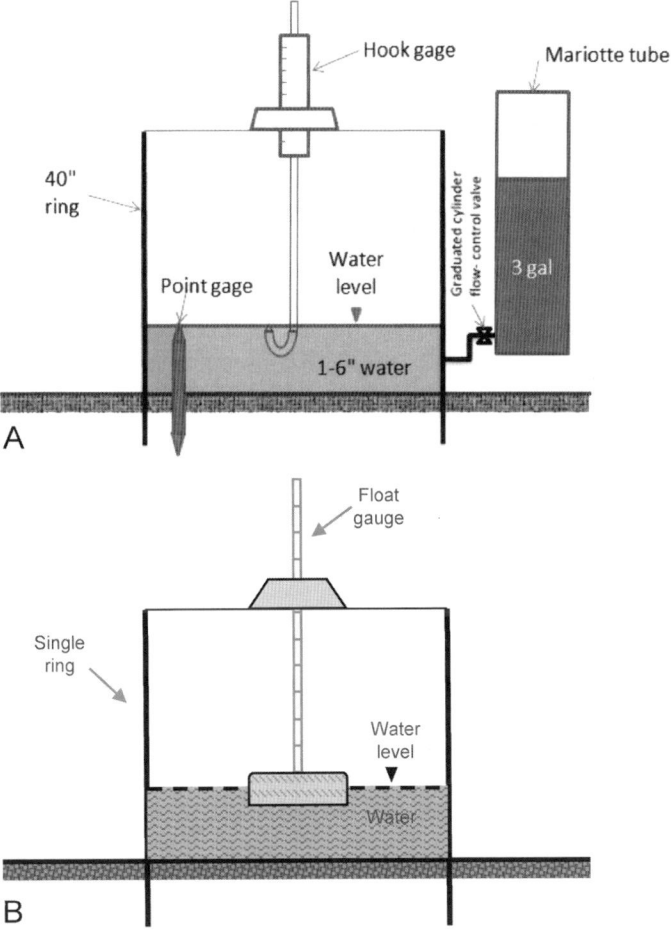

Fig. 7.5. Schematic sketch of a single-ring infiltrometer setup for the constant head method (A) (Source: Riverside County Flood Control and Water Conservation District 2011, Appendix A, page 9, modif.) and for the falling head method (B).

Material required (for one measurement)

- single ring infiltrometer
- installation tool or wooden bar
- large (plastic) hammer
- spade, large knife
- stopwatch
- graduated beaker, water bucket
- water supply container (capacity 20-100 l depending on ring size, soil texture, and initial soil water content)
- water supply and control setup for non stationary, falling-head conditions consisting of:
  - measurement bridge with float assembly (alternatively: ruler)
  - water supply container with hose and valve (alternatively: graduated beaker)
- water supply and water level control for stationary, constant head conditions consisting of:
  - water level control system for water supply based on the principle of a Mariotte bottle, either automatic (water level sensor) or manual/visual (ruler)

Preparing a site for a single-ring infiltration measurement

Basically, site preparation depends on whether the soil surface of interest is vegetated or bare. Disturbances, such as compaction and smearing but also loosening of the soil during preparation of the experiment should be avoided. In case of vegetation cover, it is recommended to cut the vegetation close to the surface by a pair of scissors. Under bare conditions, silting of the surface must be avoided, because this may limit its infiltration capacity considerably. Rocks and roots in the soil may complicate the installation of the ring considerably. If possible, these should be removed without disturbing the soil surface substantially. If removal is not feasible, a more suitable measurement site should be chosen. After driving the ring into the soil, the soil surface within the ring can be covered with a thin layer of coarse sand to avoid surface sealing during ponding.

Conducting a single-ring infiltration experiment

1. Install the single ring horizontally and without jamming it by pressing it gently, but with force into the soil manually or with a wooden bar and a plastic hammer. Problems can be alleviated by pushing down a spade or a knife along the desired perimeter of the ring, which allows the ring to be inserted more easily. The typical depth of the lower end of the ring is approximately 5 cm below the soil surface.

2. Soil surfaces prone to silting are covered with a layer of coarse sand of ca. 1 cm thickness.

3. Prepare the water supply and the gauges for reading the water levels.

4. Water is added by pouring water into the ring (bucket, graduated cylinder) or by opening the valve at the water container to achieve the desired maximum water level (usually 5 or 10 cm above the soil surface) and the measurement is started.

5. Manual or automated recording of the time-dependent water level or volume of infiltrated water until a constant infiltration rate is reached.

For falling head measurements, the water levels after equal intervals of time, or the time required for a particular change in water level are recorded. The minimum height of the water column should not be less than 5 cm during the entire measurement period. Multiple infiltration measurements may be carried out after replenishing the water in the ring.

For constant head measurements, the infiltrated amount of water is immediately determined as a function of time by recording the amount of water infiltrated per time unit. The amount of time necessary for reaching a constant infiltration rate may range from a few minutes (sandy substrates) to many hours, even days (dry, clay-rich substrates).

Data evaluation

First, the cumulative infiltration as the total amount of water infiltrated into the soil within a defined time period will be calcualted and plotted. Therefore, Eq. 7.1 and Eq. 7.2 will be used for the constant head and falling head method respectively.

$$I_s = \frac{V_w}{A\,t} \qquad (7.1)$$

$$I_s = \frac{H_w}{t} \qquad (7.2)$$

where I_s is the infiltration rate measured in the single-ring infiltrometer (mm s^{-1}), A is the base area of the single-ring infiltrometer (m^2), t is the infiltration time (s), V_w is the volume of infiltrated water (m^3), and H_w is the height change of water column in the single-ring infiltrometer (m).

b. After calculation of the infiltration rate, the rate should be plotted versus time.

7.3.1.1.2. Double-Ring Infiltrometer

Double-ring infiltrometers are considered as the accepted standard tool for measuring soil infiltration (ASTM D3385, DIN 19682-7, Blume et al. 2011). Double-ring infiltrometers prevent the lateral flow of water infiltrated from the inner ring almost completely (Fig. 7.6), resulting in a more 1D water flow, and therefore, accurate determination of the amount of water infiltrated across the area of the inner ring (Durner 2011).

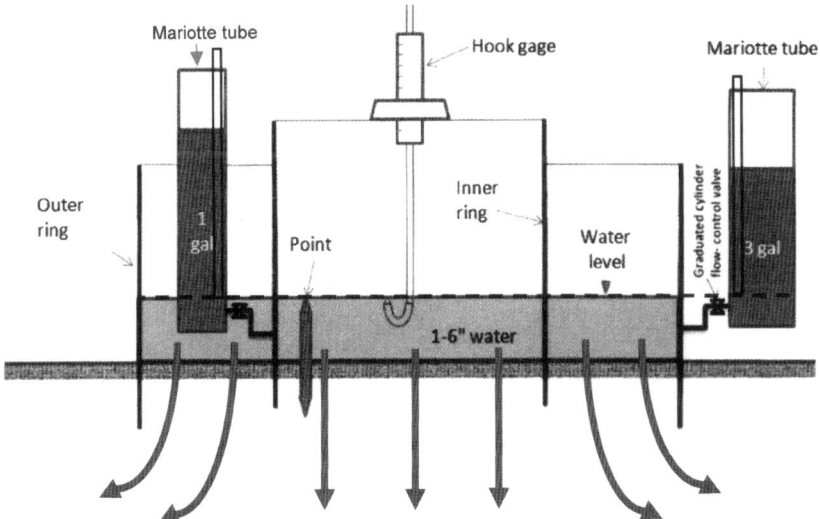

Fig. 7.6. Schematic sketch of a double-ring infiltrometer setup for constant head method (Source: Riverside County Flood Control and Water Conservation District 2011, Appendix A, page 15, modif.).

Ring size

In order to minimize the impact of the lateral water flow below the infiltrometer, the diameter of the outer ring should be considerable greater than that of the inner ring. According to DIN 19682-7 and ASTM D3385-09 the rings should have diameters of 30 and 60 cm (corresponding to 12 and 24 inches), respectively. Youngs (1987) studied the importance of ring size for infiltration measurements and reported comparable result regarding the ring sizes (cf. Fig. 7.4). Because the amount of water required for a measurement, often exceeds the available water which can be carried on the field, smaller double-ring infiltrometers are commonly also used. A standard double-ring infiltration, conforming to DIN usually includes ring pairs with 28/53, 30/55, or 32/57 cm (Fig. 7.7 left). The diameters of the three ring pairs listed here exclusively serve easy portability. Verbist et al. (2013) used a double- ring infiltrometer with rings sizes of 28 and 50 cm diameter. But ring sets of 18/32 cm also generally give reliable results (cf. Fig. 7.4) (Youngs 1987). Gregory et al. (2005) also reported compareable results from ASTM standard double-ring infiltrometers (30/60 cm) with those measured by much smaller ring sizes of 15/30 cm. Another advantage of smaller double-ring setups is the direct connection of both rings leading to one single instrument (see Fig. 7.7 right).

Fig. 7.7. Double-ring infiltrometers with inner rings diameters 28-32 cm and outer rings diameters 53-57 cm (left side, Eijkelkamp 2015) and smaller double-ring with connected rings of diameters 18/32 cm (right side).

Water supply

As for the single-ring infiltrometer, there are (the same) two procedures to supply water to double-ring infiltrometer, the constant head method and the falling head method (see section 7.3.1.1.1). To apply the constant head method correctly in the double ring infiltrometer, the water level needs to be held constant both in the inner and the outer ring (two Mariotte bottles needed) (Fig. 7.6). Because dealing with two bottles under field conditions is tedious, water may also be added to the outer ring manually.

Material required (for one measurement)
- double-ring infiltrometer
- installation tool/wooden bar
- large plastic hammer

- spade, large knife
- stopwatch
- graduated beaker, water bucket
- data sheet (see Annex 1)
- water supply container (capacity 20-100 l depending on ring size, soil texture, and initial soil water content)
- water supply and control setup for non stationary, falling-head measurements consisting of:
  - measurement bridge with float assembly and measurement setup (alternatively: ruler)
  - water supply container with hose and valve (alternatively: graduated beaker)
- water supply and water level control for stationary, constant-head measurements
  - water level control system for water supply based on the principle of a Mariotte bottle, either automatic (water level sensor) or manual/visual (ruler) water level control

Preparation of the soil surface

The soil surface must be prepared exactly in the same way as for a single-ring measurement (see section 7.3.1.1.1).

Conducting a double-ring infiltration experiment

1. According to DIN 19682-7, the first step is to install the outer ring by pressing it into the soil manually or using a wooden bar and a plastic hammer. In a second step, the inner ring is installed, in the center of the outer ring. In practical application, installing the inner ring before the outer ring has shown to be beneficial. Cutting along the perimeter of the rings with a spade or a knife may simplify ring installation. The installation depth of the rings should be 5 cm.

2. Surfaces prone to silting must be covered with a layer of coarse sand (1 cm thickness).

3. Prepare the water supply for both the inner and the outer rings and set up rulers to measure the water level in the inner ring (measuring ring).

4. First fill the outer ring, so that the water infiltrating downwards creates a buffer region which prevents water from the inner ring from flowing off radially below the bottom of the inner ring (Fig. 7.8).

5. Water is supplied to the inner ring (measuring ring) either manually or by opening the valve at the water supply tank until the maximum ponding level (generally 5 or 10 cm) has been reached. After, the actual measurement starts, take care, that the water level in the inner and outer ring will be the same during the entire experiment.

Fig. 7.8. Lateral flow of water below a double-ring infiltrometer (Source: Eijkelkamp).

6. Manual or automated recording of the amount of infiltrated water as a function of time until a constant rate of infiltration (steady state) has been reached (see section 7.3.1.1.1.).

Data evaluation

First, the cumulative infiltration recorded from the innner ring as the total amount of water infiltrated into the soil within a defined time period will be calcualted and plotted. For the analysis the same method as described for the single-ring infiltrometer will be used (Eq. 7.3 and Eq. 7.4) for the constant head and falling head method, respectively. Note, that the area A is the base area of the inner ring only, V_w is the volumen of water infiltrated ion the inner ring only, and H_w is the changes in height in the innner ring only if a double-ring infiltrometer is used.

$$I_D = \frac{V_w}{A\ t} \qquad (7.3)$$

$$I_D = \frac{H_w}{t} \qquad (7.4)$$

Special notes

Erroneous measurements may occur also using a double-ring by leaky Marriotte bottles (for the inner ring), or leaks between inner and outer rings. Measurements on sloping surfaces often promote errornous measurements because of lateral water flow from the region of the outer ring, particularly where compacted layers of soil exist close to the surface. Such layers occur widely (solifluidal layers) in low mountain ranges.

7.3.1.2. Tension infiltration

Tension infiltrometers allow for measuring infiltration capacities and unsaturated hydraulic conductivities (Reynolds 2006). In contrast to the methods described in prior sections, which operate with constant or variable head, tension infiltrometry operates at a negative pressure relative to the ambient atmospheric pressure. While methods relying on a ponding water always integrated of the entire pore spectrum of the infiltrated soil, tension infiltrometry allows to determine infiltration rates and hydraulic conductivities of different pore size ranges (Youngs 1991). This is generally achieved by carrying out measurements at different tensions (commonly 0 to -10 cm). As it it known, that the rapid drainage of macropores (draining at very low tensions close to 0) often leads to a strong drop of the infiltration rate (Schwärzel & Punzel 2007, Watson & Luxmoore 1986) very large macropores (shrinkage cracks, worm burrows, or root channels) are generally not covered by measurements using tension infiltrometers if the tension is < 0.

7.3.1.2.1. Hood Infiltrometer

Hood infiltrometers (Schwärzel & Punzel 2007) are tension infiltrometers which, compared to "normal" tension infiltrometers, have the advantage of being usable without any contact material between the instrument and the soil surface as the hood, in which the desired tension is adjusted, can be placed directly on the soil surface (Fig. 7.9). The rate of infiltration can be measured at different tensions and may be used to determine hydraulic conductivities even close to saturation.

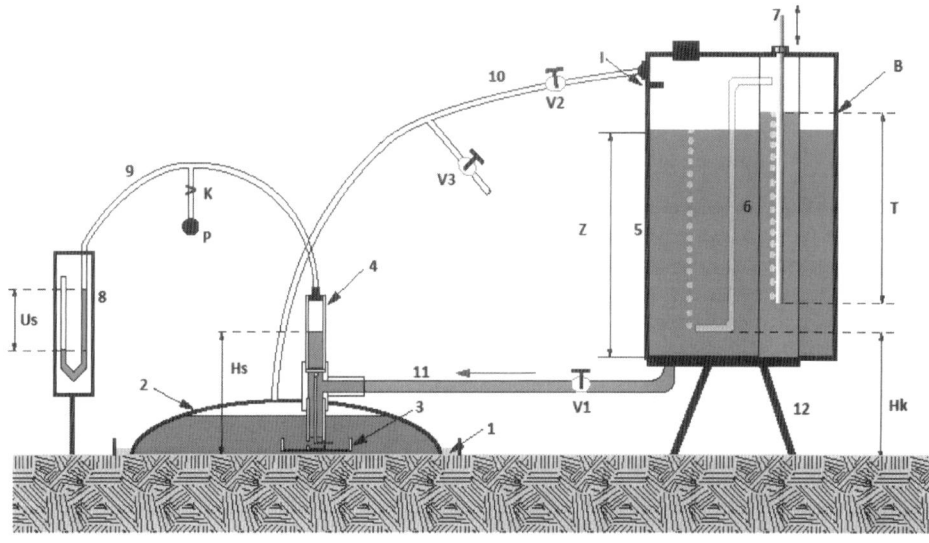

1	Fine sand sealing	B	Filling level of the bubble tower
2	Plastic hood	Hk	Distance between soil surface and air outlet
3	Buffer vessel	Hs	Water level in the standpipe
4	Standpipe	I	Maximum filling level of the infiltration reservoir
5	Infiltration reservoir	K	Valve for cutting off the pipette ball
6	Bubble Tower	P	Pipette ball
7	Air intake pipe	T	Immersion depth of the air intake pipe
8	U-tube manometer	Us	hight difference at the U-tube manometer
9	Hose connection 1	V1	Valve for separating the water volumes
10	Vent pipe	V2	Valve for separating the air volumes
11	Hose connection 2	V3	Valve for pressure adjustment
12	Stand	Z	Filling height of the infiltration reservoir

Fig. 7.9. Schematic sketch of a Hood Infiltrometer (Quelle: Umwelt-Geräte-Technik (2012).

Material required

- Hood infiltrometer (inclusive metal ring and optional pressure logger)
- watering can
- water reservoir
- water
- hammer
- spade/small shovel
- spray bottle
- stopwatch
- scissors
- data sheets and notebook

Conducting a hood infiltrometer experiment

Logging the volume of infiltrated water

The rate of infiltrated water can be measured using a stopwatch and water level markers on the water reservoir. At very low nfiltration rates, the required accuracy of a pressure logger may be too small and may require manual readings. Using an automated pressure logger is particularly recommended at high and medium infiltration rates and for successive measurements, since manual measurements are prone to inaccuracies. If a pressure logger is used, it must be calibrated prior to the actual measurement by determining the pressure within the empty reservoir (offset of the calibration line) and the pressure at a known water level (slope of the calibration line).

Preparing the measuring plot

The plot should as representative and as level as possible. Loose plant litter should be removed and any vegetation should be cut down using a scissor.

Preparing the hood infiltrometer

The U-tube manometer (Fig. 7.9) must be filled with water carefully, using a spray bottle so that no air bubbles will form. After that, the reading marker is adjusted to the zero mark, and the bubble tower must be filled with water and the tubes must be connected with all valves closed.

Running the experiment

First, gently press the metal ring into the ground so that only a few centimeters will stay outside the soil. The ring must be levelled with the ground to prevent leaks.

The water reservoir of the infiltrometer should be placed at an even spot and adjusted in a way that its bottom is oriented horizontally. The measuring hood is then placed on the measuring location, within the metal ring. Regions of the soil where roots are detectable should be avoided. To prevent leaks of the hood, and hence the entire setup, the space between hood and metal ring is filled with a sealing substrate. Fine sand or the local soil, if the clay content is not too high, serves well for this purpose. The sealing substrate is pressed tightly into the space between hood and metal ring and saturated with water from the spray bottle. After filling the main reservoir with water, the measurement can be started by opening the main valve for the water supply.

By opening the air escape valve, the hood fills with water. Pressing the hood down slightly keeps the sealing substrate from becoming dislocated. Once the hood is filled up to the marker, the air escape valve is closed. The tension at which the infiltration rate is measured equals the difference of the tensions in the U-manometer and the standpipe. The desired tension is set by the venting tube of the bubble tower. It is recommended to start measurements at low tensions, increasing them slowly until the so called "bubble point" is reached. For determining unsaturated hydraulic conductivities, measurements at two or more tensions are necessary. To fit a nonlinear trend of unsaturated hydraulic conductivity to soil water tension close to water saturation, measurements at at least three different tensions are required. If infiltration is stationary, the rate of infiltration ($\Delta z/\Delta t$) may be measured manually or using a logger. The applied tension must be checked during the measurement period to keep it stable. After the end of the measurement, the main valve of the water reservoir has to be closed, which brings the hood again to atmospheric pressure.

Data evaluation

Measured outflow rates can be related to infiltration rates by applying a factor, which depends on the hood size (Umwelt-Geräte-Technik GmbH 2012). Using the individual infiltration rates at each tension, the unsaturated hydraulic conductivities can be computed, using the analytical solution by Wooding (1968) for stationary infiltration into the soil with

$$\alpha = \frac{\ln(\frac{\Delta z1}{\Delta t1} / \frac{\Delta z2}{\Delta t2})}{h1 - h2} \tag{7.5}$$

$$K_u(h) = \frac{v_b}{(1 + \frac{4}{\pi \alpha r})} \tag{7.6}$$

$$v_b = q * \frac{dz}{dt} \tag{7.7}$$

where K_u is the unsaturated hydraulic conductivity (cm min^{-1}), h is the applied tension (suction head) during the measurement (cm), v_b is the recorded infiltration rate (cm min^{-1}), q is the hood specific dimensionless factor converting outflow rates to the infiltration rates (–), r is the radius of the infiltration area (hood radius) (cm), and α is the Gardner coefficient (cm^{-1}), which denotes the dependence of hydraulic conductivity versus tension and may be determined by measuring at two different tensions (h_1, h_2).

Special notes

Hood infiltrometers are well suited for quickly measuring hydraulic conductivities close to water saturation without using a contact material, and hence, under natural soil surface conditions. Hood infiltrometers are well suitable for field use, whereby the transportation of the instrument over large distances in the field is problematic to its size. At high infiltration rates or for large hoods, water consumption during measurements may be problematic, as as only limited water supply will be available at the field.

Measurement problems generally occur at very high infiltration rates (e.g., in forest soils), which exceed the infiltration rates at which the hood can be replenished. In case of very low infiltration rates (e.g., clay-rich soils) inaccuracies result when pressure loggers are used. This is particularly the case, when winds cause the infiltrometer to jiggle. Furthermore, setting the tension and simultaneously observing the U-tube manometer and the standpipe at high infiltration rates demands very rapid work of the operator. Common practical problems during operation are open valves or small leaks of the settled hood. These problems can mostly be avoided by carefully selecting suitable measurement locations or by sealing leaks with soil material.

7.3.1.2.2. Tension-disc infiltrometer

As the general handling of tension-disc infiltrometers (Perroux and White 1988) is the same as for hood infiltrometers (see section 7.3.1.2.1.), only some additional remarks are made here for the specific handling of the tension-disc infiltrometers. A general sketch of a tension infiltrometer is shown in Fig. 7.10.

The fundamental technical difference to a hood infiltrometers is, that the contact of the instrument with the soil is ensured by the use of a nylon membrane, which, ideally, requires a complete flat surface. The flat surface is obtained by preparing the soil surface or by the use of a contact material (Mc Kenzie et al. 2002). In general, the contact layer material used must have a higher hydraulic conductivity than the soil to be infiltrated (Schwärzel & Punzel 2007, Angulo-Jaramillo et al. 2000). Most commonly, fine sand is used as a contact and leveling material for measurements at low tensions. For measurements at higher tensions using finer grained silty

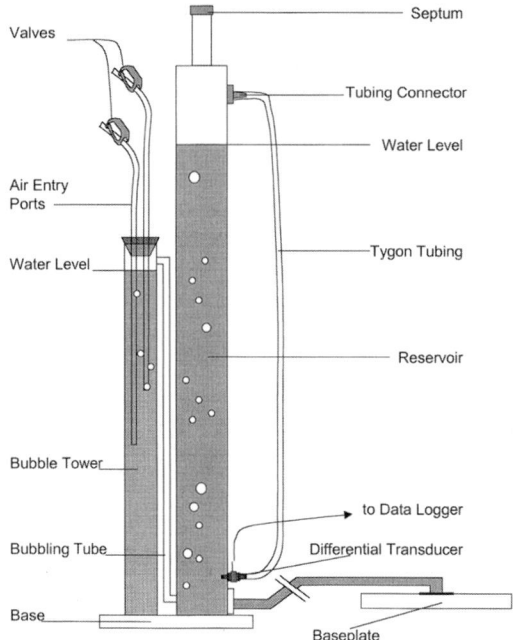

Fig. 7.10. Sketch of a tension-disc infiltrometer (Eijkelkamp 2010).

material is recommended (Durner et al. 1999, Torfs 2008). Furthermore, the porous membrane must be free of mechanical damage. The contact layer may also cause underestimating hydraulic conductivities, particularly by reducing infiltration into macropores (Reynolds et al. 2000).

Prior to the use of a tension-disc infiltrometer, its membrane (if necessary with the relevant frame) must be watered to achieve full water saturation of the membrane system.

7.3.1.3. Classifying infiltration values

The measured infiltration capacity of soils is known to depend on numerous instrumental and soil conditions. The data in Table 7.1 list commonly observed steady-state final infiltration rates for different devices or reported from different studies. As a general trend, infiltration rates decrease from coarse (sandy) to fine grained substrates (clay). Additionally, the spread of reported infiltration rates within a single soil textural class may sometimes still be considerable.

From the assignment of actual infiltration rates (see Table 7.1 and 7.2), according to Kretzschmar (1994), it becomes evident that clay-rich soils usually fall into the categories low or very low, while sandy soils are classified at least into the medium infiltration class.

Unfortunately, the correlation between soil texture and infiltrability is not as clear as shown above. For example, Rahmati et al. (2018) nicely showed, that on a global perspective, the infiltrability cannot be explained by soil texture alone and that other factors such as soil management and land use might be equally important. Therefore, it is strongly recommended to report also the current land management and land use (and if available also past land management and land use) along with the infiltration data.

In order to 'buffer' (infiltrate) intense rain events with precipitation rates of >10 mm h^{-1}, which occur occasionally during storm events, the soil infiltration should at least fall into the class 'low-medium'. Nevertheless, even soils with medium to high infiltration rates (e.g., sandy loam and sand) might fail to infiltrate extreme rainfall events, especially if short in duration, or if the soil is already fully saturated.

Such situations will led to extreme flooding events as observed e.g., in Germany during the Christmas-flood of 1993 (Meuser and Worreschk 1994), the "January-flood" in 1995 or the "Autumn-flood" in 1998 on the Mosel and Rhine rivers (Fell 1999) but also to flooding at many other locations globally.

Table 7.1. Reported infiltration rates for different instruments and from different studies.

Author	Eijkelkamp	Kretzschmar	Kretzschmar	USDA–NRCS	DEQ	Johnson	Free et al.	Sacramento county code
Year	2015	1994	1994		2016	1963	1940	
Device	double-ring	double-ring	double-ring			tubes	tubes	
Position	surface	surface	surface			surface	surface	
Unit	(mm h^{-1})	(mm h^{-1})	(mm h^{-1})	(mm h^{-1})	[mm h^{-1}]	(mm h^{-1})	(mm h^{-1})	(mm h^{-1})
Sand	>30	50	25–250	>20	102–508	not reported	not reported	25–32
Sandy loam	20–30	25	12–75	10–20	50–95	10–50	22	15–19
Loam	10–20	12	8–20	5–10	34–49	13	6	14
Clay loam	5–10	8	2.5–15	1–5	25–13	10–100	7	6
Clay	1–5	5	1.25–10	<1	<13	0–1	0–6.5	3

Table 7.2. Infiltration classes after Kretzschmar (1994).

Infiltration class	Infiltration rate (mm h^{-1})
very low	<1
low	1–5
low to medium	6–20
medium	21–63
medium to high	64–127
high	128–254
very high	>254

Fig. 7.11. Double-box slope infiltrometer with 3 separate inner boxes (6x18cm) and outer box frame (30 x 30 cm) (Dep. of Soil Science, University Trier).

7.3.1.4. Futher methods to determine infiltration rates (overview)

- Double-ring with sealed inner ring (ASTM D5093-15)
- Turf-Tec double-ring (Turf-Tec International 2014)
- Ponded infiltrometer (Prieksat et al. 1992)
- Twin-ring infiltrometer (Cook 2002)
- Mini-disc infiltrometer (Decagon Devices, Inc. 2014)
- Automated mini-disc infiltrometer (Klipa et al. 2015)
- Double-box slope infiltrometer (see Fig. 7.11, not published)
- Rainfall simulators (e.g., Iserloh et al. 2012)

7.3.2. Water Conductivity Measurements

After water has entered the soil surface (infiltration), further transport of this water within the soil is controlled by the so called hydraulic conductivity. It is an intrinsic soil property and may vary considerably vertically and laterally. The soil hydraulic conductivity is thereby a function of the primary and secondary pore systems and depends, among other factors, on texture, relevant processes of soil evolution (e.g., soil structure), and also on anthropogenic factors (e.g., deep tillage or compaction).

To assess the soil hydraulic properties, the saturated hydraulic conductivity, (K_s), the permeability rate, (P_r), or the soil matrix flux potential, (Φ), can be used. Irrespectively of the usefullness of these parameters, little is known about their quantification: i) at short- and long-term periods; ii) during different seasons (winter, summer…), or human activities (harvesting, tilling…); iii) under different climatic conditions or soil types; iv) different land use and management; and, v) its connectivity mechanisms at intra-plot scale (Rodrigo Comino et al. 2016a). Moreover, several authors stated that field measurements of these properties differ compared to standard laboratory techniques (e.g., Kumar et al. 2010, Rienzner and Gandolfi 2014). Additionally, several problems are reported for field experiments measuring these properties, especially at slopy conditions or when a high fraction of stones, cracks, or macropores are abound (e.g., Bodhinayake and Cheng Si 2004, Buczko et al. 2006, Rodrigo Comino et al. 2016b).

In the following, two popular methods for measuring the saturated hydraulic conductivity under field conditions will be introduced.

7.3.2.1. Guelph-Permeameter

The Guelph-permeameter (GP) can be considered as one of the most useful tools to measure the soil permeability (Fig. 7.12). Through a modified Mariotte bottle device, the Guelph-permeameter imposes a constant water level in a borehole discharging the water three-dimensionally into the surrounding soil, which can be easily performed by a single person (Elrick et al. 1989, Elrick and Reynolds 1992, Reynolds and Elrick 1987, Reynolds and Lewis 2012). This instrument allows measuring the soil permeability in all soil types and requires only about 2.5 l of water, whereby the measurements can be obtained for soil depth raning from 15 to 75 cm depth below surface.

Material required

- Guelph-Permeameter
- Soil auger
- Sizing auger
- Water container (2.5 l water for each measurement)
- Little bucket for water to fill the Guelph-permeameter
- Stopwatch
- Folding rule
- Shovel
- Plastic bags
- Data sheets (see Annex 2)

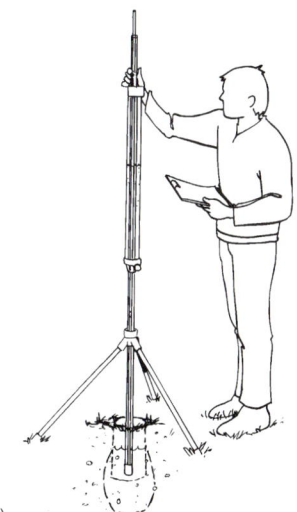

Fig. 7.12. Sketch of a Guelph-permeameter measurement (Eijkelkamp).

Conducting a Guelph-permeameter experiment

Each measurement in the field will be carried out following the procedure illustrated in Fig. 7.13 by Rodrigo Comino et al. (2016a, 2016b) and listed in the following:

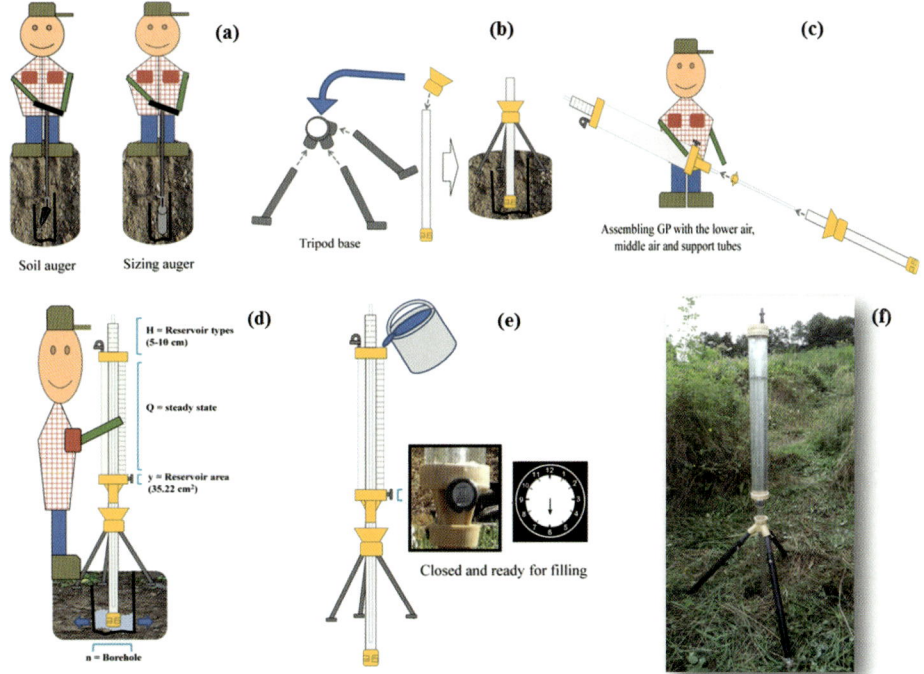

Fig. 7.13. Illustration of a Guelph-permeameter measurement in the field according to Rodrigo Comino et al. (2016a, 2016b).

1. Carefully complete the Guelph-Permeameter data sheet providing information on the experimental plot.
2. Prepare the borehole (Fig. 7.13 a): Here, is recommended to use a sharp auger to avoid smearing effects at the borehole wall, which will affect the hydraulic properties and always turn the auger in the same direction. By using a brush, the borehole may be emptied of remaining soil materials. Take a subsample of the augered material and store it in a plastic bag to determine the starting (initial) water content of the soil in the laboratory.
3. Measure the final radius and depth of the borehole (cm).
4. Assemble the Guelph-permeameter: (b) Setup the tripod and fix the legs to avoid movement (Fig. 7.13 b) with the wide end oriented up, over the outside of the support (Fig. 7.13 b). Connect the lower air tube to the middle air tube at the reservoir base and connect the support tube over the air tube (connect it firmly into the adapter in the bottom of the reservoir base) (Fig. 7.13 c). Set the support tube into the tripod, and ensure that the well height indicator is in place before, and connect the upper air tube to the middle air tube (Fig. 7.13 d). Take care that the opening valve is closed when filling the reservoir (Fig. 7.13 e) with ~2.5 l of water and finally start the measurement by opening the valve to the reservoir.

Experimental procedure

When the Guelph-permeameter is fully filled with water, level with the bubble scale and start the experiment by turning the valve on. The infiltration measurement will start, when the level in the bubble tower remains constant and first bubbles from the air inlet are detectable. In general, repeated measurements in the same borehole can be performed to archive reliable results. Total duration for each run is about 30 minutes (or it can be stopped when the main water reservoir is emptied). Over the course of the measurement, the water level must be checked every 2 minutes, (use a stopwatch). After finishing the measurements, soil samples should be collected (0.3 to 0.5 kg in plastic bags) for laboratory analysis of soil organic carbon content, gravel (> 2 mm) content, and soil texture (sand, silt, clay content), as well as final water content.

Data evaluation

Experiments performed with the Guelph-permeameter should be performed in triplicates at the same site (with similar soil conditions) (at least three repetitions under similar environmental plot characteristics are recommended). Based on the replicates, the mean steady state infiltration rates (cm min^{-1}), based on the 2 minute readings, can be calculated.

Finally, simple analytical equations can be used to calculate three fundamental hydrological parameters: the permeability rate (P_r), the saturated hydraulic conductivity (K_s), and the matrix flux potential (Φ) (Reynolds 1986, Reynolds and Elrick 2002, Rodrigo Comino et al. 2016b, Zhang et al. 1998).

$$P_r = \left(\frac{\frac{y*Q1}{1000}}{2\pi(a*H1)}\right) * 60 \ (\text{mm h}^{-1}) \tag{7.8}$$

$$K_s = \left(\frac{C1*Q1}{2\pi H1^2 + \pi a^2 C1 + 2\pi\left(\frac{H1}{a}\right)}\right) * 600 \ (\text{mm h}^{-1}) \tag{7.9}$$

$$\Phi = \left(\frac{C1*Q1}{(2\pi H1^2 + \pi a^2 C1)\alpha + 2\pi(H1)}\right) * 6000 \ (\text{mm}^2 \text{ h}^{-1}) \tag{7.10}$$

where $C1$ is a shape factor (-), Y is the area of the combined reservoir (cm^2) (for most commercial sytems $y = 35.22$ cm^2), $Q1$ is the quasi steady-state flow rate out of the permeameter and into the soil (cm min^{-1}), $H1$ is the pressure head (ponding height) of water established in the borehole (cm), α is the macroscopic capillary lenght parameter, which depends on the soil texture-structure category (and can be looked up from tables), and a is the borehole radius (cm).

Special notes

Measurement errors may be related to: i) incorrect assembly of the air tubes; ii) insufficient preparation of the borehole (e.g., cracks, leftover of fine material, insufficient depth; and, iii) incorrectly reading of the water level).

Additionally, it has to be noted here, that the combination of the Guelph-permeameter with other measurements, such as rainfall simulator measurements for permeability, infiltration, runoff, and soil erosion assessment performed at the same location are rarely performed as stated in literature (Gupta et al. 2006 1993, Leonard and Andrieux 1998, Rodrigo Comino et al. 2016a). The same also hold for other infiltration measurements (sing-ring, double-ring, tension-disc, or the borehole method).

Further, high variability of the recorded infiltration results has been reported for repeated infiltration measurements using the Guelpg-permeameter, but also for other infiltration techniques, for a single experimental plot. A reason for that can be the inner plot heterogeneity of stones, micro-topographical changes, shrinkage cracks, biological macropores, or variabilities in soil texture or structure, or differences in soil tillage, which all impact the soil hydraulic conductivity.

The impact of small scale heterogeneous pedological characteristics and human influences on the observed hydrological parameters (infiltrability, saturated hydraulic conductivity, matrix flux potential) has been observed in various studies (e.g., Follain et al. 2012, Govers 1985, Govers et al. 1994, Imeson and Lavee 1998, Quiquerez et al. 2014, Ruiz Sinoga and Martinez Murillo 2009). Therefore, it is necessary to account for soil and land use specific differences to cover the information at the landscape level (e.g., erosion, colluvium, different managed agricultural units). Especially, the distribution of coarse fragments in the soil profile or landscape might impact the yhdrological response (e.g., infiltration) substantially, and therefore, will also impact the partitioning (infiltration and runoff) of incoming water (e.g., precipitation, irrigation). For example, vineyards require repeated soil management, and those soil tillage operations can impact the hydraulic properties substantially.

7.3.2.2. Borehole-method

Basically, the borehole-method measures the rise of the groundwater level in a drained borehole under defined conditions. To do this, a borehole is drilled into the soil to a depth below the groundwater level with as little disturbance of the soil as possible and the rising water table will be monitored and used to calculate the saturated hydraulic conductivity, K_s, of the deep soil.

Material required

- Pürckhauer auger, hammer, and extraction tool (if necessary also an extension); alternatively use an Edelman auger (if necessary also an extension) tool
- Bailer
- Measuring tape with (redirecting) holder and float
- Filter pipe, usually 1 m long, diameter appropriate to the auger (if necessary also an extension)
- Stopwatch
- Writing materials

Conducting a borehole measurement

In a first step, the borehole will be drilled up to a depth below the groundwater table. In a next step, the borehole is bailed out, a float with fixed tape is placed on the water surface and the rate of groundwater table rise in the borehole is measured. Based on a set of several measurements of the groundwater rise and by the help of some computation, the saturated hydraulic conductivity of the soil around the borehole and between borehole bottom and reference groundwater level can be calculated.

Such in-situ measurements of the saturated hydraulic conductivity below the groundwater level, according to DIN 19682-8, requires only simple instruments, and is therefore easy to perform and provides valueable information due to:

- taking undisturbed soil samples in saturated soils for laboratory analyses is always difficult, time consuming, and often impossible, but not required for the borehole method.
- transport of undisturbed saturated samples always alters their intrinsic soil structure inside the cores.

Furthermore, results of laboratory analyses reflect real field conditions only rarely, whereas borehole measurements characterize a much larger and undisturbed part of the soil by its integrated measurement.

Here, it has to be noted, that the borehole-method assumes isotropic distribution of the saturated hydraulic conductivity, K_s, with same K_s in vertical and horizontal direction.

If previous soil mapping indicates anisotropic or heterogenous soils (layering in the groundwater influenced layers), this can be by handled by:

- repeated borehole measurements above and below a layer boundary and use of an adapted computational code or by use of the piezometer method decribed by Kirkham (1945).
- increase the number of measurement points (locations) if strong heterogeneities in soil texture (vertical, horizontal or nested) will be observed.

Before conducting a borehole measurement, it is necessary to perform a soil profile description to a depth of at least 2 meters to answer following questions:

Q 1: Is there just ONE or are there SEVERAL layers with possibly different K_s? If more than one, note depth of the boundary between the layers!

Q 2: Is there an impermeable layer closer than 50 cm to the lowest permeable layer? If so, at what depth does it start?

Q 3: Is the profile representative of the site? If not, increase number of measuring locations!

Q 4: Is the soil clayic or does the soil exhibit dominant silt and clay fractions? If so, document any layering. Please be aware that clayic subsoils might have extremely low K_s values of 0.01 to 0.50 cm d^{-1}

Q 5: Is it a peat soil? If so, be aware of peat type and structure, as compressed peat fibers are horizontally permeable but quite impermeable vertically.

In general, the boreholes should be prepared a few days prior to measurements, especially in clayic soils to get sufficient time for the level to stabilize and to reduce smearing effects of the borehole walls.

Drill one or more boreholes with auger of 8 cm down to a minimum of 20 cm below the actual groundwater level. If necessary, stabilize the borehole wall by a filter screen especially for sandy soils. Check borehole depth by the use of the auger as a "measuring tape", mount the tape holder above the borehole, and prepare the float.

After equilibration of the groundwater level the following measurements according to Fig. 7.14 should be performed:

M1   Record the depth of the groundwater level in the borehole (cm)

M2	Record the distance between the borehole bottom and the next considerable less permeable layer (cm)
DATE	Provide date for the measurement (year, month, and day)
TEXT	Provide unambiguous name of the measuring point
r	Record the radius of the borehole

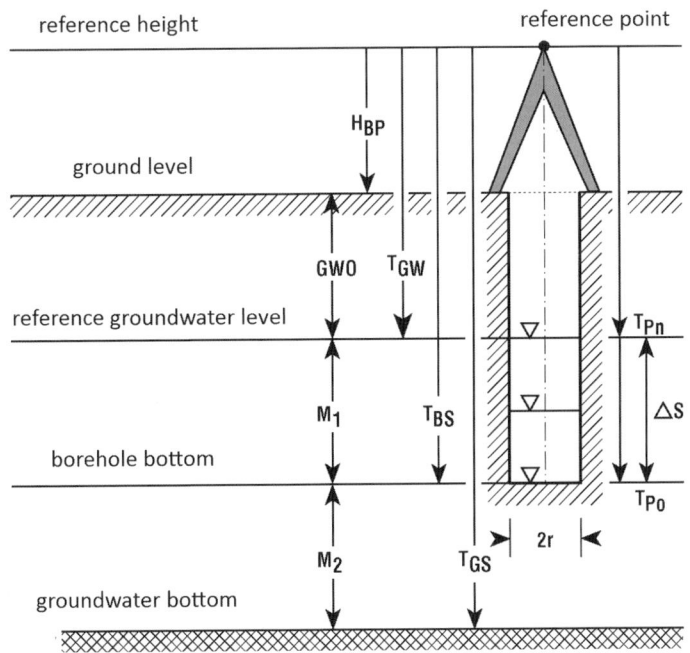

Fig. 7.14. Sketch of the borehole method. Explanation given in Table 7.3.

Table 7.3. Registration sheet for borehole measurements, with main data to be recorded. Explanation for Fig. 7.14 and Fig. 7.15.

Parameter[#]	Meaning	Registration (example)
DATE	year, month, day of measurement [yyyy-mm-dd]	10-06-2019
TEXT	name of measuring point	borehole-Geseke_10_06_2019
NO	number of measurements (at least 5)	11
R	radius of borehole in cm	4
HBP	reference height above ground level in cm	34
TGW	depth of reference groundwater level below reference height (cm)	100
TBS	depth of borehole bottom below reference height in (cm)	183
TGS	depth of groundwater bottom below reference height (cm)	200
GWO	reference groundwater level (GWO = TGW - HBP)	
M1	depth of the water column in the borehole (M1 = TBS - TGW)	
M2	distance between borehole bottom and next considerably less permeable layer (M2 = TGS - TBS)	

[#] to be used in pseudo-code shown in fig. 7.15

154   Soil water parameters

HBP   Record the height above ground level (cm)
TGW   Record the reference groundwater level below reference height (cm)
TBS   Record the borehole bottom below reference height (cm)
TGS   Record the groundwater bottom below reference height (cm)

After, the measurement can be started (please keep the stopwatch ready).

- remove 20 to 40 cm water from borehole (this is usually done with a bailer, but take care and do not deepen the borehole while removing water)
- quickly lower float on new groundwater level
- start stopwatch and
- read the float depth

Check whether constant time intervals in between 5 to 30 seconds or constant intervals of water level rise in between 1 to 3 cm are more appropriate for the current measuring situation, but never change measurement type during one measurement series. Compile all data according to Table 7.4a or Table 7.4b.

Table 7.4. Registration sheet for series of borehole measurements.
a. Depth of the groundwater level for fixed time intervals.
b. Time when rising groundwater level has reached a fixed depth interval.

Measurement	Time (s)	Depth (cm)	Measurement	Time (s)	Depth (cm)
1	0	180.0	1	0	180
2	15	179.3	2	…	179
3	30	178.5	3	…	178
4	45	177.9	4	…	177
5	60	177.3	5	…	176
6	90	176	6	…	175
7	120	174.6	7	…	174
8	180	172.1	8	…	173
9	240	169.7	9	…	172
10	300	167.4	10	…	
11	360	165.3	11		

An indicator when the original groundwater level is reached is given by slowdown of the increase of the groundwater level (larger time intervals for a given unit of increasing water level). As this slower incease will lead to a non-linear function of the water-level/time relationship only the first 25% of the recorded data will be used for calculating the saturated hydraulic conductivity and the rest (75%) of the data will be neglected. Note, that these percentages are only a rough estimate and the actual threshold can be determined by plotting the results (time versus groundwater level) and selecting only the linear part of the curve. Finally, it should be checked whether different depths of water table after draining the borehole yield different initial groundwater level increase and select a based on these findings determine the optimal starting groundwater level. Perfom at least 5 repeated measurements under equal conditions to assure comparability. From experiments it has been shown that using 11 repetions of same measurements will yield reliable K_s data (see also evaluation example below).

Data evaluation

To ensure a straight forward calculation of the saturated hydraulic conductivity from repeated measurements in the same borehole the procedure as sketched in the pseudo-code in Fig. 7.15 should be followed. This pseudo-code can be translated in any programming language and can be checked against the example values und example evaluation provided below.

```
read from database       TEXT           # name of measurement point
read from database       DATE           # year, month, day of measurement (yyyy-mm-dd)
read from database       HBP            # reference height above groundwater level (cm)
read from database       TGW            # depth of reference groundwater level below reference height (cm)
read from database       TBS            # depth of borehole bottom below reference height (cm)
read from database       TGS            # depth of groundwater bottom below reference height (cm)
read from database       R              # radius of borehole (cm)
read from database       NO             # number of measurments (at least 5)
read from database       Time [i] TP [i]  # time and depth of groundwater level at time step i, with i = 1 to i = ANZ

# Calculations and Initializing
                GWO = TGW - HBP
                M1 = TBS - TGW
                M2 = TGS - TBS
                n = (NO - 1) * NO / 2

                if (M2 >= 0.5 * M1)
                    radq = 400000 * R²
                    f_1 = 20
                    return = "full-space"
                else
                    radq = 360000 * R²
                    f_1 = 10
                    return = "full-space"
                end

        kf_summe = 0
        kf_summeq = 0

# Check for plausibility
                if {(NO<5) or (NO>11)}
                    return = "Number of measurements less than 5 or more than 11"
                end

                for i = 1 to i = NO-1
                        if (TP[i]<=TP[i + 1])
                            return = "Time step to small"

                        if {(M1<20) or (M1>200)}
                            return = "Distance between reference groundwater level and borehole bottom
                                      less than 20 cm or more than 200 cm"

                        if (M2<15)
                            return = "Distance between bottom and groundwater bottom less than 15 cm"
                        end
                end

# Calculations
                for i to k <NO
                        kf_mit[k] = 0
                        delta_s = TP[k] -TP[i + 1]
                        sm = TP[k] - TGW - delta_s / 2
                        delta_t = Time[i + 1] - Time [k]
                        kf[i,k] = radq / ((M1 + f_1 *radq) * (2-sm / M1) * sm)) * (delta_s / delta_t)
                        kf_summe = kf_summe + kf[i,k]
                        kf_summeq = kf_summeq + kf[i,k] * kf[i,k]
                        kf_mit[k] = kf_mit[k] + kf[i,k]
                        kf_summeq = sqrt{kf_summeq - (kf_summe + kf_summe) / n} / (n-1)}
                        kf[i,k] = kf_summe / n
```

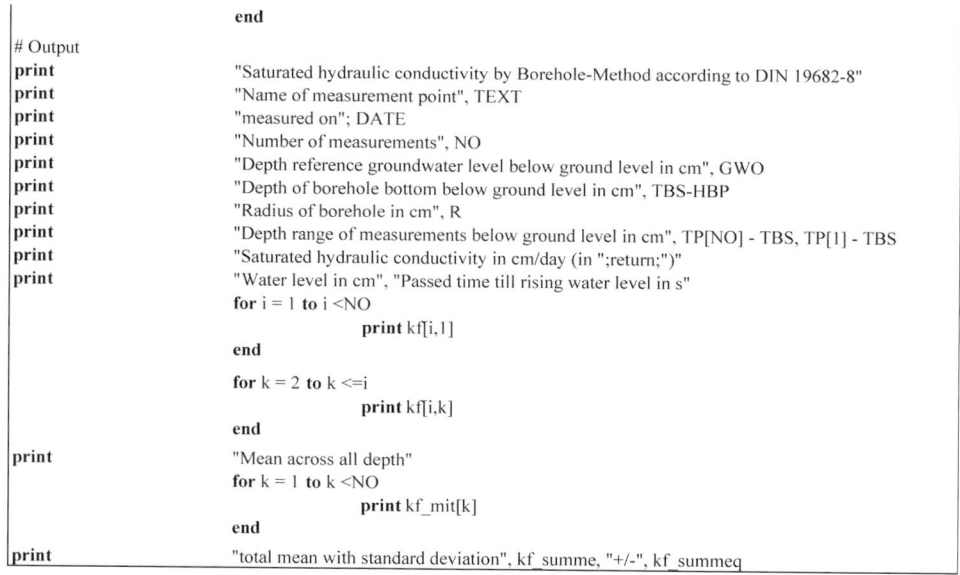

Fig. 7.15. Computation of the saturated hydraulic conductivity from borehole measurements. Given in pseudo-code; explanation of input data are provided in Table 7.4.

Evaluation example

Saturated water conductivity by Borehole-Method following DIN 19682-8	
Name of measuring point	Bohrloch – Geseke
Measured on	10. June 1987
Number of measurements	11
Depth reference groundwater level below ground level (cm)	66
Depth of borehole bottom below ground level (cm)	149
Radius of borehole (cm)	4
Depth range of measurements below ground level (cm)	131 to 146

In the following example, the evaluation of the data of the example values listed in Table 7.3 and Table 7.4a is shown, leading to the calculated K_s values listed in Table 7.4b. Hereby, each water level as combined with all possible groundwater levels and time intervals were used.

With respect to Table 7.6 the calculated K_s values presented in Table 7.5 show an average saturated hydraulic conductivity with a very low variability.

Table 7.5. Saturated hydraulic conductivity, K_s, in cm d^{-1}.

Water level (cm)	Passed time till rising to water level (seconds)								
	0	15	30	45	60	90	120	180	240	360
180.0										
179.3	26.4									
178.5	28.3	30.2								
177.9	26.4	26.4	22.6							
177.3	25.5	25.2	22.7	22.7						
176.0	25.2	24.9	23.6	23.9	24.6					
174.6	25.5	25.4	24.6	25.0	25.6	26.6				
172.1	24.9	24.8	24.3	24.5	24.7	24.7	23.8			
169.7	24.4	24.3	23.9	24.0	24.1	24.1	23.4	23.1		
167.4	24.0	23.9	23.5	23.6	23.7	23.5	23.1	22.7	22.3	
165.3	23.4	23.3	23.0	23.0	23.0	22.9	22.4	22.0	21.4	20.6
mean across all depths	25.4	25.4	23.5	24.3	24.3	24.4	23.2	22.6	21.9	20.6

Total mean with standard deviation: 24.2 ± 1.6

Table 7.6. General ranking of K_s values.

K_s values (cm d^{-1})	Permeability class
≤ 5	low
5 to ≤ 40	average
40 to ≤ 100	high
> 100	very high

7.3.2.3. Further methods (overview)

- Open end tests and packer tests in boreholes/USBR 7310-89 (US Department of the Interior 1990a)
- Well permeameter test/USBR 7300-89 (US Department of the Interior 1990b).
- Aardvark permeameter (Theron et a. 2010)
- Amoozemeter/Compact Constant Head permeameter (Amoozegar 1989, 1992)

7.4. References

Amoozegar, A., 1989. A compact constant-head permeameter for measuring saturated hydraulic conductivity of the vadose zone. Soil Sci. Soc. J. 53: 1356–1361.

Amoozegar, A., 1992. Compact constant head permeameter: A convenient device for measuring hydraulic conductivity. In: Topp, C.G. et al. (ed.), Advances in measurement of soil physical properties: bringing theory into practice. Soil Sci. Soc. Am. Spec Publ. No. 30, Madison. USA.

Angulo-Jaramillo, R., Vandervaere, J.-P., Roulier, S., Thony, J.-L., Gaudet, J.-P., Vauclin, M., 2000. Field measurement of soil surface hydraulic properties by disc and ring infiltrometers – A review and recent developments. Soil & Tillage Research 55: 1–29. DOI: 10.1016/S0167-1987(00)00098-2

ASTM D3385-09, Standard test method for infiltration rate of soils in field using double-ring infiltrometer, ASTM International, West Conshohocken, PA, 2009, United States.

ASTM D5093-15. Standard Test Method for Field Measurement of Infiltration Rate Using Double-Ring Infiltrometer with Sealed-Inner Ring, ASTM International, West Conshohocken, PA, 2015.

ASTM D5126. 1998. Standard guide for comparison of field methods for determining hydraulic conductivity in the vadose zone. 100 Barr Harbor Drive, PO Box C700, West Conshohocken, PA 19428-2959, United States.

ASTM D6391. 2006. Standard test method for field measurement of hydraulic conductivity limits of porous materials using two stages of infiltration from a borehole. 100 Barr Harbor Drive, PO Box C700, West Conshohocken, PA 19428-2959, United States.

Blume, H.P., Stahr, K., Leinweber, P., 2011. Bodenkundliches Praktikum. Spektrum Akademischer Verlag, Heidelberg, Germany. ISBN 978-3-8274-1553

Bodhinayake, W., Cheng Si, B., 2004. Near-saturated surface soil hydraulic properties under different land uses in the St Denis National Wildlife Area, Saskatchewan, Canada. Hydrol. Process. 18: 2835–2850. DOI: 10.1002/hyp.1497

Buczko, U., Bens, O., Hüttl, R.F., 2006. Tillage effects on hydraulic properties and macroporosity in silty and sandy soils. Soil Sci. Soc. Am. J. 70: doi: 10.2136/sssaj2006.0046

Chapuis, R.P., 2007. References on field permeability tests performed in boreholes and monitoring wells. Rapport technique EPM RT 2007-04. 18 pages. École Polytechnique. Montréal. Canada.

Clothier, B.E., 1991. Infiltration. In: Smith, K.A., Mullins, C.E. (ed.), Soil and environmental analysis. Physical methods, p. 239–280. 2nd ed., rev. and expanded. ISBN 0-8247-0414-2.

Cook, F.J., 2002. The twin-ring method for measuring saturated hydraulic conductivity and sorptivity in the field. In: McKenzie, N.J. et al., Soil Physical Mesaurement and Interpretation for Land Evaluation: 90–107, Collingwood, Australia.

Decagon Devices, Inc., 2014: Mini Disk Infiltrometer. Pullman, USA.

DEQ 2016, Standard Percolation Test. https://www.deq.idaho.gov/media/60178064/tgm-2-3-0316.pdf

Di Prima, S., Lassabatere, L., Bagarello, V., Iovino, M., Angulo-Jaramillo, R., 2016. Testing a new automated single ring infiltrometer for Beerkan infiltration experiments. Geoderma 262: 20–34. DOI: 10.1016/j.geoderma.2015.08.006

DIN 19682-7:2015-08. 2015. Bodenbeschaffenheit – Felduntersuchungen – Teil 7: Bestimmung der Infiltrationsrate mit dem Doppelring-Infiltrometer.

DIN 19682-8. Bodenbeschaffenheit – Felduntersuchungen – Teil 8: Bestimmung der Wasserdurchlässigkeit mit der Bohrlochmethode.

Dirksen, C., 1991. Unsaturated Hydraulic Conductivity. p 183–238. In: Smith, K.A., Mullins, C.E. (ed.), Soil and environmental analysis. Physical methods. 2nd ed., rev. and expanded. ISBN 0-8247-0414-2.

Eijkelkamp. 2010. Tension infiltrometer. Giesbeek.

Elrick, D.E., Reynolds, W.D., 1992. Methods for analyzing constant-head well Permeameter data. Soil Sci. Soc. Am. J. 56: 320–323. DOI: 10.2136/sssaj1992.03615995005600010052x

Elrick, D.E., Reynolds, W.D., Tan, K.A., 1989. Hydraulic conductivity measurements in the unsaturated zone using improved well analyses. Ground Water Monit. Remediat. 9: 184–193. DOI:10.1111/j.1745-6592.1989.tb01162.x

Fohrer, N., Bormann, H., Miegel, K., Casper, M., Bronstert, A., Schumann, A., Weiler, M., 2016. Hydrologie. UTB basics, Stuttgart, Germany. ISBN 9783825245139

Follain, S., Ciampalini, R., Crabit, A., Coulouma, G., Garnier, F., 2012. Effects of redistribution processes on rock fragment variability within a vineyard topsoil in Mediterranean France. Geomorphology 175–176: 45–53. DOI: 10.1016/j.geomorph.2012.06.017

Free, G.R., Browning, G.M., Musgrave, G.W., 1940. Relative Infiltration and Related Physical Characteristics of Certain Soils. United States Department of Agriculture. Technical Bulletin, 729

Govers, G., 1985. Selectivity and transport capacity of thin flows in relation to rill erosion. Catena 12: 35–49. DOI: 10.1016/S0341-8162(85)80003-5

Govers, G., Vandaele, K., Desmet, P., Poesen, J., Bunte, K., 1994. The role of tillage in soil redistribution on hillslopes. Eur. J. Soil Sci. 45: 469–478. DOI: 10.1111/j.1365-2389.1994.tb00532.x

Gregory, J.H., Dukes, M.D., Miller, G.L., Jones, P.H., 2005. Analysis of double-ring infiltration techniques and development of a simple automatic water delivery system. Applied Turfgrass Science, 2: 1–7.

Gupta, R.K., Rudra, R.P., Dickinson, W.T., Patni, N.K., Wall, G.J., 1993. Comparison of saturated hydraulic conductivity measured by various field methods. Trans. ASAE 36, 51–55. DOI: 10.13031/2013.28313

Gupta, R.K., Rudra, R.P., Parkin, G., 2006. Analysis of spatial variability of hydraulic conductivity at field scale. Can. Biosyst. Eng. 48: 155–162.

Imeson, A.C., Lavee, H., 1998. Soil erosion and climate change: the transect approach and the influence of scale. Geomorphology 23: 219–227. DOI: 10.1016/S0169-555X(98)00005-1

Iserloh, T., Ries, J.B., Cerdà, A., Echeverría, M.T., Fister, W., Geißler, C., Kuhn, N.J., León, F.J., Peters, P., Schindewolf, M., Schmidt, J., Scholten, T., Seeger, M., 2012. Comparative measurements with seven rainfall simulators on uniform bare fallow land. Zeitschrift für Geomorphologie, Suppl. 57, 1: 11–26.

Johnson, A.I., 1963. A field method for measurement of infiltration. Geological Survey Water, 1544 (F), pp. 1–17. Washington, United States.

Kirkham, D., 1945. Proposed Method for Field Measurement of Permeability of Soil Below the Water Table. Proceedings Soil Science Society of America, Vol. 10, 1945, pp. 58-68.

Klipa, V., Snehota, M., Dohnal, M., 2015. New automatic minidisk infiltrometer: design and testing. J. Hydrol. Hydromech. 63: 110–116.

Kumar, S., Sekhar, M., Reddy, D.V., Mohan Kumar, M.S., 2010. Estimation of soil hydraulic properties and their uncertainty: comparison between laboratory and field experiment. Hydrol. Process. 24, 3426–3435. DOI: 10.1002/hyp.7775

Kuosa, H., Niemeläinen, E., Korkealaakso, J., 2014. Pervious pavement testing methods. State-of-the-Art and laboratory and field guideline for performance assessment. Research Report VTT-R-08225-13, Espoo, Finland.

Kretzschmar 1994. Kulturtechnisch-Bodenkundliches Praktikum – Ausgewählte Labor- und Feldmethoden.

Leonard, J., Andrieux, P., 1998. Infiltration characteristics of soils in Mediterranean vineyards in Southern France. Catena 32: 209–223. DOI: 10.1016/S0341-8162(98)00049-6

Maniak, U., 2010. Hydrologie und Wasserwirtschaft – Eine Einführung für Ingenieure. Springer, Heidelberg, Germany. ISBN 978-3-642-05396-2

Mc Kenzie, Kenzie, N.J., Cresswell, H.P., Green, T.W., 2002. Field Measurement of Unsaturated Hydraulic Conductivity Using Tension Infiltrometers. In: Mc Kenzie, N.J., Coughlan, K. CResswell, H.P. (ed.): Soil Physical Measurements and Interpretation for Land Evaluation. Collingwood. Australia.

Mertens, J., Jacques, D., Vanderborght, J., Feyen, J., 2002. Characterisation of the field-saturated hydraulic conductivity on a hillslope: In situ ring pressure infiltrometer measurements. Journal of Hydrology, 263: 217–229. DOI: 10.1016/S0022-1694(02)00052-5

Oosterbaan, R.J., Ritzema, H.P., 1992. Hooghoudt"s drainage equation adjusted for entrance resistance and sloping land. 5th International Drainage Workshop, Water and Power Development Authority and International Commission on Irrigation and Drainage, At Lahore, Pakistan, Volume: In: Vlotman, W.F. (Ed.), Proceedings, Vol. II: p: 2–18 – 2–28.

Perroux, K.M., White, I., 1988. Design for disc permeameters. Soil Sci. Soc. Am. J. 52: 1205–1215.

Prieksat, M.A., Ankeny, M.D., Kasper, T.C., 1992. Design for an automated, self regulating, single ring infiltrometer. Soil Sci. Soc. Am. J. 56: 1409–1411.

Quiquerez, A., Chevigny, E., Allemand, P., Curmi, P., Petit, C., Grandjean, P., 2014. Assessing the impact of soil surface characteristics on vineyard erosion from very high spatial resolution aerial images (Côte de Beaune, Burgundy, France). Catena 116: 163–172. DOI: 10.1016/j.catena.2013.12.002

Rahmati, M., Weihermüller, L. et al., 2018. Development and analysis of the Soil Water Infiltration Global database. Earth Syst. Sci. Data, 10(3): 1237–1263.

Reynolds, W.D., 2006. Tension infiltrometer measurements: Implications of pressure head offset due to contact sand. Vadose Zone Journal 5: 1287–1292.

Reynolds, W.D., 1986. The Guelph Permeameter method for in situ measurement of field-saturated hydraulic conductivity and matrix flux potential. Thesis, University of Guelph, Guelph, Ontario, Canada.

Reynolds, W.D., Bowman, B.T., Brunke, R.R., Drury, C.F., Tan, C.S., 2000. Comparison of tension infiltrometer, pressure infiltrometer, and soil core estimates of saturated hydraulic conductivity. Soil Sci. Soc. Am. J. 64: 478–484.

Reynolds, W.D., Elrick, D.E., 2002. Constant head well permeameter (vadose zone). In: Dane, J.H., Topp, G.C. (Eds.), Methods of Soil Analysis, Physical Methods. Soil Science Society of America, Inc. pp. 844–858, Madison, WI, United States.

Reynolds, W.D., Elrick, D.E., 1987. A laboratory and numerical assessment of the Guelph permeameter method. Soil Sci. 144: 282–299.

Reynolds, W.D., Lewis, J.K., 2012. A drive point application of the Guelph Permeameter method for coarse-textured soils. Geoderma 187–188: 59–66. DOI: 10.1016/j.geoderma.2012.04.004

Rienzner, M., Gandolfi, C., 2014. Investigation of spatial and temporal variability of saturated soil hydraulic conductivity at the field-scale. Soil Tillage Res. 135: 28–40. DOI: 10.1016/j.still.2013.08.012

Riverside county Flood Control and Water Conservation District, 2011. Design Handbook for Low Impact Development Best Management Practices – Appendix A – Infiltration Testing. Riverside.

Rodrigo Comino, J., Ruiz Sinoga, J.D., Senciales González, J.M., Guerra-Merchán, A., Seeger, M., Ries, J.B., 2016a. High variability of soil erosion and hydrological processes in Mediterranean hillslope vineyards (Montes de Málaga, Spain). Catena 145: 274–284. DOI:10.1016/j.catena.2016.06.012

Rodrigo Comino, J., Seeger, M., Senciales, J.M., Ruiz-Sinoga, J.D., Ries, J.B., 2016b. Variación espacio-temporal de los procesos hidrológicos del suelo en viñedos con elevadas pendientes (Valle del Ruwer-Mosela, Alemania). Cuad. Investig. Geográfica 42: 281–306. DOI: 10.18172/cig.2934

Ruiz Sinoga, J.D., Martinez Murillo, J.F., 2009. Effects of soil surface components on soil hydrological behaviour in a dry Mediterranean environment (Southern Spain). Geomorphology 108: 234–245. DOI:10.1016/j.geomorph.2009.01.012

Sacramento County Code. Agricultural activities and water use and conservation. https://qcode.us/codes/sacramentocounty/view.php?topic=14

Schwärzel, K., Punzel, J., 2007. Hood infiltrometer – a new type of tension infiltrometer. Soil Sci. Soc. Am. J. 71: 1438–1447.

TGL 31222-06. 1980-09. Physikalische Bodenuntersuchung – Bestimmung des Durchlässigkeitsbeiwertes nach der Bohrlochmethode.

Theron, E., Le Roux, P.A.L., Hensley, M., Van Rensburg, L., 2010. Evaluation of the Aardvark constant head soil permeameter to predict saturated hydraulic conductivity. WIT Transactions on Ecology and Environment 134: 153–162.

Torfs, S., 2008. Evaluation of field methods to determine hydraulic properties of stony soils in arid zones of Chile. MSc Thesis, Univ. Gent, Gent.

Turf-Tec International. 2014: Turf-Tec Infiltrometer – Easy to use Double Ring Infiltrometer. www.turf.tec.com/IN2lit.html

Umwelt-Geräte-Technik GmbH (2012) Manual Hood Infiltrometer IL-2700. Issue: 30/01/12. Müncheberg, Germany.

USDA-NRCS. Soil Infiltration. https://www.nrcs.usda.gov/Internet/FSE_DOCUMENTS/nrcs142p2_053268.pdf

US Department of the Interior, 1990a. Earth Manual. Part 2. A Water Resources Technical Publication. Procedure for constant-head hydraulic conductivity tests in single drill holes (USBR 7310-89).

US Department of the Interior, 1990b. Earth Manual. Part 2. A Water Resources Technical Publication. Procedure for Performing Field Permeability Testing by the Well Permeameter Method (USBR 7300-89).

Verbist, K.M.J., Cornelis, W.M., Torfs, S., Gabriels, D., 2013. Comparing methods to determine hydraulic conductivities on stony soils. Soil Sci. Soc. Am. J. 77: 25–42.

Vereecken, H., L. Weihermüller, S. Assouline, J. Šimůnek, A. Verhoef, M. Herbst, N. Archer, B. Mohanty, C. Montzka, J. Vanderborght, G. Balsamo, M. Bechtold, A. Boone, S. Chadburn, M. Cuntz, B. Decharme, A. Ducharne, M. Ek, S. Garrigues, K. Goergen, J. Ingwersen, S. Kollet, D.M. Lawrence, Q. Li, D. Or, S. Swenson, P. de Vrese, R. Walko, Y. Wu, Y. Xue, 2019. Infiltration from the Pedon to Global Grid Scales: An Overview and Outlook for Land Surface Modeling. Vadose Zone Journal 18(1): doi: 10.2136/vzj2018.10.0191

Watson, K.W., Luxmoore, R.J., 1986. Estimating macroporosity in a forest watershed by use of a tension infiltrometer. Soil Sci. Soc. Am. J. 50: 578–582.

Winter, F., 2013. Prozessorientierte Modellierung der Abflussbildung und -konzentration auf verschlämmungsgefährdeten landwirtschaftlichen Nutzflächen. Diss. Universität der Bundeswehr, Munich, Germany. 978-3-8440-2242-1

Youngs, E.G., 1987. Estimating hydraulic conductivity values from ring infiltrometer measurements. European Journal of Soil Sciences 38: 623–632. doi: 10.1111/j.1365-2389.1987.tb02159.x

Youngs, E.G., 1991. Hydraulic Conductivity of Saturated Soils. In: Smith K.A., Mullins, C.E. (eds.): Soil and environmental analysis: Physical methods, p. 141–182. 2nd ed., rev. and expanded. ISBN 0-8247-0414-2.

Zhang, Z.F., Groenevelt, P.H., Parkin, G.W., 1998. The well-shape factor for the measurement of soil hydraulic properties using the Guelph Permeameter. Soil Tillage Res. 49: 219–221. DOI:10.1016/S0167-1987(98)00174-3

Annex:

Annex 1: Field data sheet for Double ring infiltrometer measurements (modified from Eijkelkamp 2015).

Date:
Location:
Remarks:

Field Protocol for Double Ring Infiltrometer Measurements

A	B		C	D	E	F	G	H
Time	Water Level		Cumulative Time	Time Interval	Infiltration	Infiltration capacity	Infiltration capacity	Cumulative Infiltration
	before filling	after filling						
Reading (hours, min)	Reading (mm)	Reading (mm)	Determine from A (min)	Determine from (min)	Determine from (mm)	Calculate from D & E (mm/min)	Calculate from F ../...	Determine from E (mm)
			start = 0					start = 0

Annex 2: Example of a data sheet for Guelph permeameter measurements.

Study area ( ); Plot ( ); Number of measurements ( ); Date ( ); Label ( )

Investigator/s	Date
Study area description	Coordinates
	Height (m.a.s.l.)
	Slope inclination (°)
Depth hole (cm)	Radius (cm)
Reservoir used during test: combined ( ) or inner only ( )	
Soil water content at start (%)	Organic matter content (%)
Gravel content (%)	Fine material (%)

Water level in well (5 cm)		*Water level in well (10 cm)*	
Time	Water level in reservoir	Time	Water level in reservoir
(min)	(cm)	(min)	(cm)
00:00		00:00	

Observations
Soil tillage (ploughing etc)
Stone cover, stoniness, cracks
Biological activity (insects, roots, small animals)
Vegetation cover
Problems during measurement

Soil mechanical parameters

8. Experimental field methods to quantify soil erosion by water and wind-driven rain

Stefan Wirtz, Thomas Iserloh, Miriam Marzen, Wolfgang Fister

8.1. Introduction
8.1.1. General introduction

Erosion by water is the most widespread mechanism of soil erosion on all five continents (Miehlich 2003, Oldeman et al. 1991, Richter 1998). The processes causing erosion by water are the subject of numerous textbooks (e.g., Auerswald 1998, Morgan 2005, Ries 2011, Roth 1995, Schwertmann et al. 1990). Key erosion events are caused by intense rainfall of high intensity (>10 mm h^{-1}), which impacts on a sparsely vegetated to vegetation free soil surface (Ries 2011). Despite a long history of research and countless studies, there is no worldwide comprehensive database of rates of soil erosion (García-Ruiz et al. 2015, Ries 2011). Empirical work and research into the fundamental processes is required to understand these processes in detail, to enlarge the database, and to allow modeling the processes controlling soil erosion by water (Iserloh et al. 2013c, Ries 2010, Ries et al. 2013, Stroosnijder 2005).

Erosion research distinguishes between splash erosion (Fig. 8.1C), interrill erosion (Fig. 8.1D), rill erosion (Fig. 8.1A, B), and gully erosion processes (e.g., Richter 1998, Ries 2011). Wind may influence all erosion processes listed (Fister et al. 2012, Iserloh et al. 2013a, Marzen et al. 2015, 2016, 2017, Ries et al. 2014, Schmidt et al. 2017).

The impact of rain drops on the soil surface (splash) is the initial process, which fragments soil aggregates, removes soil particles from the soil surface, and transports them with or without the splash water, depositing soil particles radially around the point of drop impact (Auerswald 1998, de Lima 1989, Le Bissonnais 1996, van Dijk et al. 2002a, 2002b). This redistributes soil material on the soil surface, making it susceptible to other transport processes. Amount and size of the eroded particles are essentially a function of the kinetic energy and the momentum of the impacting raindrops (de Lima 1990, Dunne et al. 2010, Furbish et al. 2007, Kinnell 2005, Legout et al. 2005, Leguédois et al. 2005, Riezebos and Epema 1985). Slope, surface roughness, and particularly wind may intensify splash erosion considerably (Marzen et al. 2016). Where the rain intensity exceeds the infiltration rate or where the soil is saturated, surface runoff arises and further soil particles can be detached and transported away. Due to the laminar characteristics of the thin surface flow the actual erosion is relatively limited. Impacting raindrops increase the detachment and transport of particles significantly, due to increased turbulence when impacting on the water surface (raindrop impacted flow erosion, Kinnell 2005). In the course of this interrill erosion, small domains of runoff concentration may develop and a form of rill-interrill flow develops. Where this occurs, surface rills of 1–2 cm depth form, channeling the runoff and turning it into deeper grooves (dm-range), which may deepen further (given corresponding amounts of water) and finally form gullies (m-range; Ries 2011).

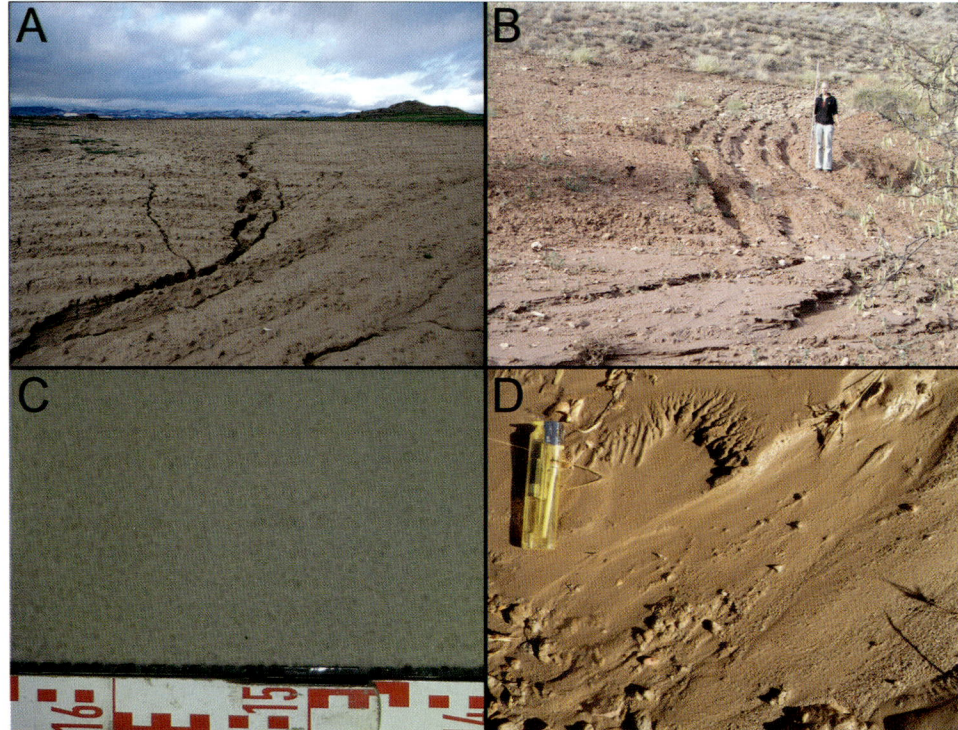

Fig. 8.1. A and B: rill erosion on arable land; C: splash, and D: interrill erosion.

Rill erosion is caused by runoff concentration. It is viewed as a process, which takes place once the energy of running water exceeds certain parameters of erodibility of a soil (Knapen et al. 2007). Linear erosion processes may be distinguished in terms of geomorphology and hydrology by acknowledging that running water is most effective as an agent of erosion and transport where it concentrates in rills (Merz and Bryan 1993). Runoff depth, delta-h, and flow velocity in rills have higher values compared to interrill runoff, and increase the forces acting on soil particles by many times than under planar runoff conditions (Poesen 1987).

Experimental soil erosion research has certain advantages compared to erosion measurement in plots under natural rain conditions, because experiments can be repeated multiple times under controlled experimental boundary conditions, which allows addressing specific questions, shortens the times of measurement, and allows to systematically produce data where they are missing (Bowyer-Bower and Burt 1989, Iserloh 2013, Stroosnijder 2005). The processes of particle detachment and transport can be studied independently of weather constraints, which is particularly handy to study arid and semi-arid regions (Martínez-Murillo et al. 2013, Ries et al. 2013). While conventional plot studies of quantitative erosion and surface runoff under conditions of intense rain require a minimum of 8–10 years of measurements to obtain acceptable statistics (Richter 1983), annual rates of precipitation of <300 mm, which often take the form of a few intense rain events, require much longer periods of observation to obtain comparably reliable data (Ries 2001). Experiments, on the other hand, are spatially unlimited in their application, while allowing to adapt individual parameters, such as duration and intensity of rain, in order to address specific problems (Bowyer-Bower and Burt 1989, Clarke and Walsh 2007, Kuhn et al. 2014).

Empirical-experimental soil erosion research comprise both laboratory and field research. Under laboratory conditions, soils of different texture may be studied under natural or artificial rain conditions in order to model specialized and abstract situations (Brunton and Bryan 2000, Bryan and Poesen 1989, Gilley et al. 1990, Huang et al. 1996, Kuhn et al. 2014, Mancilla et al. 2005, Marzen et al. 2016). The results of such tests may also be very abstract, and may be not or only very speculatively transferable to natural soils and conditions (e.g., Giménez and Govers 2002). For this reason, field tests on natural, authochtonous substrates are absolutely indispensable for soil erosion research (e.g., Ries et al. 2014).

8.1.2. Parameters to be measured, Si-units, measurement basics

Primary parameters that need to be measured in order to quantify soil erosion are the amount of eroded material (g), the amount of runoff (l) and the ratio of these quantities, and the sediment load concentration (g l^{-1}). To be able to assess these parameters, knowledge of the used volume of water (l), the rain and runoff intensity of precipitation (both in mm h^{-1} or l h^{-1}), the duration of the experiment (s), and the times of sampling (s) is required and must be recorded.

By measuring flow velocities (m s^{-1}), rill cross section (cm), water level (cm), slope angle (°), the amount of runoff, the runoff intensity (l s^{-1}) during rill experiments, further hydraulic parameters such as the cross section of flow (cm^2), wetted perimeter (cm), hydraulic radius (cm) shear stress (Pa), unit length shear force (N), stream power (W m^{-2}), unit stream power (W m^{-1}), transport rate (kg s^{-1}), and the detachment rate (kg) can be calculated.

8.1.3. Range and limits of the applicability of the experimental approach

Physical and spatial limitations of experimental setups do not allow a 1:1 simulation of natural conditions (Iserloh 2013). Thus, experiments always represent strong abstractions of the physical reality, i.e., the data so acquired are not natural. However, they afford the opportunity to focus on understanding a particular problem or process, to formulate hypotheses and to test them. In-situ experiments approach realistic conditions in so far, as they embrace and make use of the autochthonous soil volume, in particular the soil surface, unlike laboratory experiments, which are based on allochthonous, disturbed soil samples (Iserloh 2013).

The components and their combinations described below have proven to be cost-effective, functional experimental assemblies in numerous experiments. Modifications are always possible and a desired feature in the absence of a standard procedure. Applying new ideas and experience and better funding to allow modifications, allow improvements to the experimental setups. The description below thus documents the current state of things.

8.2. Selecting the most appropriate method

- Splash, interrill erosion, surface runoff, and infiltration may be studied with the help of a small portable rainfall simulator setup or a portable wind and rainfall simulator (PWRS). Due to the complicated design of the latter, the number of possible repeat measurements in a given period of time is much smaller compared to those possible with a small portable rainfall simulator. This allows a larger number of sites to be studied and compared using a small portable rainfall simulator. The portable wind and rainfall simulator, in contrast, provides a more truthful model of natural conditions, because of its larger plot size. Additionally, it allows for additional parameters to be acquired.

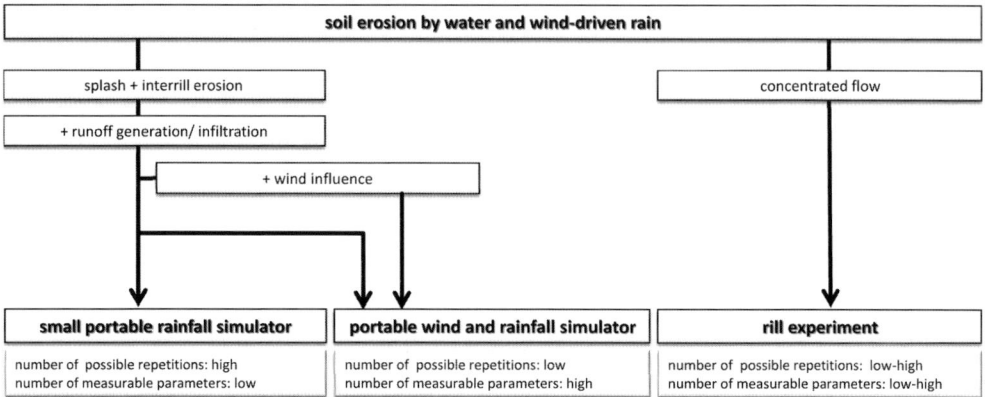

Fig. 8.2. Decision tree to determine the most suitable experimental method to quantify soil erosion by water- and wind-driven rain.

- Where the influence of wind on a soil surface or on the parameters listed previously is to be assessed, the portable wind and rainfall simulator is the method of choice. Using a small portable rainfall simulator is not possible for this purpose, because its plastic canvas prevents the wind from exerting any effect on the processes to be studied. The canvas makes individual experiments simpler to conduct and better comparable, but precludes studying wind, or wind-influenced soil erosion.
- Where effects of concentrated runoff along predefined pathways are to be studied, a rill experiment is called for. Depending on the particular question at hand, the number of parameters acquired during the experiment may be increased or decreased, this allows conducting, depending on the measurement effort, a larger or smaller number of repeat measurements.

Interpretation and assessment of measured parameters

The interpretation of experimental data has shown, that it makes sense to define erosion threat classes. The figures listed in Table 8.1 are based on empirical values obtained by rainfall simulation and rill experiments.

Table 8.1. Erosion risk classification scheme based on runoff coefficient, mass of eroded material, and sediment concentration (load) (modified from Wirtz et al. 2012a)

Risk class	Risk	Runoff-coefficient (%)	Mass of eroded material (g)	Sediment load concentration (g l^{-1})
5	extreme	>75	>40	>10
4	very high	50.1–75	8.1–40	8.1–10
3	high	30.1–50	4.1–8	6.1–8
2	medium	10.1–30	1.1–4	4.1–6
1	low	1–10	0.1–1	2.1–4
0	very low	<1	<0.1	<2

8.3. Experimental methods

8.3.1. Small portable rainfall simulator

8.3.1.1. Required equipment and preparation of an experiment

- Small irrigation setup:

Aluminum frame, nozzle holder with nozzle, reinforced plastic canvas to protect setup from evaporation and wind, plot, water barrel, flow meter, rod to hold the flow meter, bilge pump with battery and battery charger, hose connecting pump and flow meter, host connecting flow meter and nozzle, calibration plate.

- Tools:

Geometer (yardstick), rubber hammer to drive in plot borders, small shovel, screwdrivers (normal and phillips), spade, pick ax, level (to align the irrigation setup horizontally and vertically), pipe wrench.

- Other requirements:

Sampling bag for disturbed soil samples for measuring soil water content and grain size distribution, method to measure surface roughness, clinometer including a compass (to measure plot inclination and orientation), camera (for documentation), a plumb bob (on a string), water canister, wide-neck plastic bottles (250, 500 ml), graduated beakers, board with chalk, data recording sheets, silicon band, stopwatch, waterproof pen, pegs, cable binders.

8.3.1.2. Instructions for set up and carrying out an experiment

- **Installing the plot.** The plot delimiter is laid out on the chosen soil surface and the runoff-plate is oriented so as to point downhill. The plot delimiter is carefully driven into the

Fig. 8.3. Small portable rainfall simulator.

soil with a rubber hammer. Care must be taken to modify the soil surface by this as little as possible. Where soil surfaces are crusted and dry, wetting them along the margin of the plot helps to minimize destruction as the delimiter is driven into the soil. In order to position the collection containers (plastic bottle and graduated beakers) optimally, a little hole is dug underneath the runoff-shield.

- **Mapping and photographing the soil surface.** Once the plot boundary is in place, the plot surface must be mapped in detail. To that end the position, slope angle, orientation, the surface roughness and the areal percentage of vegetation cover and rock surface are determined and documented. The position is obtained by GPS and the slope angle is measured with a clinometer. Surface roughness is measured with the method of Saleh (1993). The percentage of vegetation and rocks covering the plot is estimated. Location-specific factors, e.g., silting of the topsoil or mud cracks in it are documented and recorded. The plot surface is photographed before and after each simulation; the camera must be positioned at right angle over the plot. For identification purposes a little slate board with the identification number, location and a 'b' or an 'a' (denoting **b**efore or **a**fter irrigation) is placed to the right of the plot boundary and photographed with the plot. Yardsticks are placed to the left and below the plot for scale. This arrangement is kept constant for all photographs. Where irrigation plots are directly exposed to the sun, two photos are taken: one with the surface exposed to the sun, and another with the surface shaded.

- **Setting up the rainfall simulator.** After mapping, photographing, and documenting the location, the aluminum framework is assembled. The nozzle is inspected for dirt and cleaned if necessary and the wind protection tarp is carefully pulled over the framework, leaving an opening where the water hose enters the setup. After that, the setup is oriented and fixed with pegs. Using the plumb lead, the nozzle is locked in place at the correct position over the plot and the nozzle holder is arranged vertically using a level. Correctly orienting and positioning the framework and nozzle is critically important for the quality of the data obtained later. The aluminum framework is then anchored to the ground with four pegs. The rod for the flow meter is now struck into the soil at the same height as the rainfall simulator and the flow meter is attached to it. The water barrel is set up close to the rainfall simulator and is filled with water. The hoses connecting the bilge pump with the entry side of the flow meter and a second hose connecting the exit side of the flow meter with the nozzle are installed. The bilge pump is placed into the water barrel and the electrical cable is connected to the battery. Keep both the pump switch (on the cable) and the battery away from water. Subsequently, the calibration plate is placed on the plot boundary in order to calibrate rain intensity. Now, the bilge pump is turned on and the valve is slowly opened to achieve the desired water flow (depends on desired intensity and plot size).

- **Calibration of rain intensity.** Rain intensity is calibrated by irrigating for 2.5 minutes and collecting the runoff in a calibration vessel.

- **Carrying out the actual experiment.** If the calibration was successful, the rainfall simulation experiment is started immediately thereafter without interruption. To that end, the calibration plate is removed from the plot. During irrigation, after onset of surface runoff, samples of runoff water are collected in pre-labeled and pre-weighed plastic bottles of 250 or 500 ml capacity. Depending on the amount of runoff, multiple bottles are required for a single experiment. The onset of runoff, which is the moment when surface runoff becomes visible at the surface, and every exchange of bottles are recorded. Moreover, other irregularities such as the formation of small pools or flushing processes at the soil surface are registered.

- **Post-experiment rain intensity check.** Directly following the experiment, the rain intensity is re-checked for 2.5 minutes to ascertain that the intensity has remained constant during the experiment.

- **After the experiment.** Once the plastic tarp and the aluminum setup have been removed, the plot with its margin delimiters still in place, is photographed again. Subsequently, two topsoil samples are taken; one from the irrigated surface *within* the plot and dry sample from a similar surface *outside* the plot. These samples are used to determine grain size distribution and soil water content later in the laboratory.

8.3.1.3. Interpreting the experimental results

Runoff, sediment and dissolved load are determined in every one of the collected bottles. The contents of bottles from the same measurement interval are added later. Furthermore, the soil water content and the grain size distributions of the soil samples taken from the plot are measured. In order to determine surface runoff, the full wide neck bottles are weighed as soon as possible to minimize dissolution effects. From the weight differences between the weights of the empty bottles (less the sediment in them) and their tare weights, the surface runoff is computed.

The sediment load is determined gravimetrically.

The amount of dissolved load in the runoff is obtained by measuring the conductivity of the collected runoff and is measured in the field, in order to rule out changes by slowly dissolving solids. The samples are selected randomly so that all locations and conductivities are covered.

Soil samples are taken on and off the irrigated plots to determine initial and final water contents of the soils. The type of topsoil is determined from the corresponding particle size distribution, according to Emde and Szöcs (2000).

8.3.1.4. Concluding remarks on this method

There is a multitude of experimental setups. Their key differences are in the rainfall they produce (intensity, spatial distribution, drop size and velocity, kinetic energy). Plot sizes, designs and testing procedures differ also. Consequently, experiments with different experimental setups yield very different erosion rates, so that the standardization of rainfall simulation setups and experiments remains a great challenge.

8.3.2. The Portable Wind and Rainfall Simulator (PWRS)

8.3.2.1. Required equipment and preparation of PWRS-experiments

Careful assembly of a PWRS-experiment requires a few hours. The actual amount of time depends on the experience of the operator, the structure of the soil, into which the rails (Fig. 8.5) and the sediment trap (Fig. 8.6) must be embedded, and the specific surface characterization.

The setup consists of four separate collapsible plates (aluminum and perspex), which are connected and stabilized by aluminum frames. These are set up on steel respectively wooden rails, which are inserted a few centimeters into the ground. Thus stabilizing the structure and preventing an undercutting of the test plot. Wind is generated by a two-blade, axial propeller, which is able to produce wind speeds of between 3 and 8 m s^{-1}. The rotating air stream is sent through a hose-like duct (transformer), which is 4 m long, from where it is guided into a honeycomb in order to turn it into quasi laminar-flowing wind. The lower array of tubes (height 4 cm) have been sealed in order to simulate a boundary layer.

The study area can be irrigated by four downward-spraying nozzles, which produce a full-cone spray and are attached 0.7 m above the soil surface at the ceiling of the tunnel (Fig. 8.7). A generator is required to drive the electrical pump, which delivers the water from out of a barrel through a four way splitter evenly to the four nozzles and maintains a constant flux of water of about 4*60 l h^{-1} (240 l in total) at 20 kPa (0.2 bar). Depending on the type of experiment, the

Fig. 8.4. Portable Wind and Rainfall Simulator (PWRS).

Fig. 8.5. Plot of a PWRS-experiment. Note the rails on either side of the rainfall simulator which form base of the lateral perspex windows.

Fig. 8.6. Combined sediment trap.

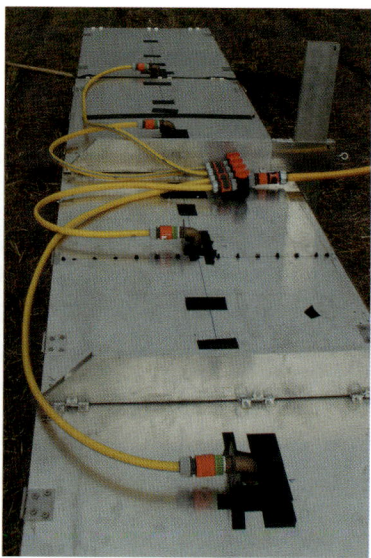

Fig. 8.7. Nozzle arrangement at the top of the tunnel.

nozzles are attached in different places, to compensate for changing irrigation intensities caused by the wind. The test area in the frontal and distal parts of the apparatus is reduced in size; to enable comparable irrigation intensities in experiments with wind (96 mm h^{-1}) and without wind (88 mm h^{-1}) (Fister et al. 2012).

The eroded material is collected in a sediment trap, which is equally able to trap sedimentary material eroded by water and eolian processes. Inappropriate installation of the assembly may cause destabilization of the rear end of the test area, which could result in undercutting by water. Continuously collected runoff data afford a simple means to check for such impairment.

8.3.2.2. Setup instructions for PWRS-field experiments

The metal rails are oriented in a straight line and parallel to each other, following the direction of steepest descent. In order to compensate for the unevenness of the surface, they are driven into the soil to a depth where all hollows are blocked to delimit the entire test area. This must be done with utmost care, because the tunnel walls and hence the irrigation tunnel are later set up on them, and control both the irrigation and the wind field. To minimize influence of a damaged soil surface (or crust) on the experiment, no sediment is collected from the marginal areas. This is achieved by leaving lateral gaps of 1 to 2 cm width on either side of the sediment trap and the rails where test field and sediment trap abut. In field experiments, a rather small slope angle is desirable, which should be constant over the length of the tunnel to allow a stable setup with constant irrigation and wind conditions along the entire length of the tunnel. Nevertheless, slope inclinations up to 25% have been tested successfully.

Accurate and surface-parallel installation of the sediment trap is very important to allow adequate measurements and to prevent water from undercutting the experimental setup.

After installing rails and sediment trap, the plot area is photographed and plot characteristics such as surface roughness, plant and stone coverage, or soil moisture content are recorded (cf. section 8.3.1.2). Because the experimental sequence is continuous, photos and soil samples (for soil water content) can only be taken either before or after an experimental sequence.

The lateral (perspex) sheets are set up vertically on the rails, connected by stabilizer profiles, starting from the beginning of the rail to its end. The first sheet is connected to the honeycomb and the transformer. The transformer is then connected to the propeller housing with a belt.

The nozzles are attached to the top plates of the tunnel and the irrigation system is set up, after having been cleaned. Before the experiment, both the gasoline and oil supply of the power generator as well as the water level in the storage tank have to be checked and replenished prior the experiment.

The channel is set up so that both the runoff and the wind are able to flow unimpeded through the tunnel.

8.3.2.3. Running an experiment

Success of any PWRS field experiment and the quality of acquired data critically depends on the careful installation of the lateral rails and sediment trap. Neither micro nor macro-relief or inclination of the plot should be extreme. They also should be as similar as possible along the entire length of the plot.

An experimental test sequence, which consists of four individual experiments on the same plot, was designed as result of a range of different test experiments. This multi-phase experimental procedure was tested intensively and its methodology optimized. The individual experiments are set up to be carried out consecutively, so that the experimental conditions of the second, third, and fourth run are harmonized by the preceding experiment. These harmonization regarding soil water content and structure of the soil surface help to correctly interpret the data in terms of the influence of wind on the erosion by water (raindrops, surface runoff). Such a consecutive setup allows both, the individual interpretation of the partial experiments and using the third and fourth experiments to assess the effects of wind-driven rain on erosion processes. The setup defines that the wind direction in the tunnel parallels the inclination of the soil surface. There are physical factors, which limit the applicability to natural rain- and storm events. A high degree of reproducibility of these experiments is obtained, because wind speed and irrigation intensity are known and highly reproducible and can also be measured and controlled during experiments.

Phase 0: Simulation of wind erosion (duration: 10 min; wind speed approx. 8 m s^{-1})
Procedure: The experiment simulates the erosion effect of wind on the experimental plot. The amount of wind-eroded material is measured to quantify the relative sensitivity of the plot surface to wind. This quantity is not relevant to the interpretation of the erosion effects of wind-driven rain.

Phase 1: Simulation of the effects of rain on a dry soil surface (duration 30 min, precipitation intensity: 96 mm h^{-1})
Procedure: During this initial rainfall simulation the raindrops impact on dry soil.
Function: With respect to the whole test sequence, phase 1 can be interpreted as a single rainfall event, which yields information on surface runoff and eroded material under conditions of heavy rain. As part of the test sequence it also serves to homogenize surface properties across the entire plot, in terms of soil water content and roughness and in terms of washing easily erodible material off the soil surface.

Phase 2: Simulation of the effects of rain on a moist soil surface. (Duration: 30 min; precipitation intensity: 96 mm h^{-1})
Procedure: During phase 2 of the experiment, raindrops impact the already moist soil surface with already modified erosion-relevant parameters, such as water conductivity, sealed pores or reduced matric potential.
Function: With respect to the whole test sequence, phase 2 is the zero sample test for the combined wind and rainfall simulation (phase 3). It may also be interpreted in terms of particular soil water content close to or at field capacity.

Phase 3: Simulation of the effects of simultaneous wind and rain on a moist soil surface (duration: 30 min.; wind speed ~8 m s^{-1}; precipitation intensity: 88 mm h^{-1})
Procedure: Both wind and rainfall simulation is activated to enable a simultaneous wind and rainfall erosion experiment.
Function: The data obtained are used to quantify the effect of wind on surface runoff and soil erosion under the influence of wind in relation to the experiment on wet surface without wind (phase 2).

8.3.2.4. Measurements during the experiment

Due to the large amount of material collected in the course of such experiments, the following procedure is suggested:

Surface runoff is determined by noting the time it takes to fill one of the 500 ml bottles, which once full, is emptied into a bucket for a period of 2.5 minutes. After 2.5 minutes, the bucket is replaced with a new (empty) one, which yields a temporal resolution of 2.5 minutes and a total of 12 intervals within the duration of the entire experiment (30 min). The buckets with runoff are allowed to settle for at least 12 hours (considered sufficient for sedimentation of the bulk of particles of all size classes). The clear supernatant is then decanted and the sediment is filled into clearly labeled PET-bottles for transport. This leads to a high time resolution and a manageable amount of sample material on one hand, but comes at the cost of a non-quantifiable loss of material, particularly of clay-size fraction, on the other. This introduces a partially systematic bias, which means that decanting water from the buckets should be done by the same person.

The content of the sediment traps is emptied into PET-bottles.

All sample bottles are consistently labeled and documented (field notebook).

The samples are filtered and dried, subsequently the dry weight of the eroded material is determined per 2.5 minute interval. The development over time and the amount of runoff is calculated from the values observed during the simulation.

8.3.2.5. Interpreting the data obtained from the simulation

Depending on the problem at hand, the data are evaluated statistically. Depending on quantity and quality of the data (e.g., normal distribution), applied statistical procedures involve descriptive statistics and (paired) comparisons of means (of runoff, amount eroded and suspended sediment concentration) to investigate differences among and between different types of simulations. The results are best visualized in bar chart diagrams and box plots. For exemplary data analysis see Iserloh (2013a) and Marzen et al. (2015).

8.3.3. Rill experiment

8.3.3.1. Equipment required and preparation of rill experiments

There are numerous variations of rill experiments. They differ in the number of parameters to be measured, the amount of water used, and inflow intensity.

Water for the experiment comes from a suitable source, which may be a source, a river or a lake, as the water is not required to have drinking water quality. The water is withdrawn from the source using a gasoline powered pump (here used and proved many times: Honda WMP 20X, see Fig. 8.10A). 2000 l of water are required for each experiment, which are withdrawn and pumped into standard water tanks of 1000 l capacity (Fig. 8.9A) and subsequently brought to the rill to be studied on a suitable vehicle. The pump is not only used to fill the water into the tanks, but also for the experiment proper.

Water transport: Where the rill to be studied is far away from a road, the water needs to be transported there using a cascade of hoses and additional plastic tanks (both 2" suction and pressure hoses are required). IBC tanks feature a rough DN 50 thread (S60x60), which necessitates

Fig. 8.8. Rill experiment.

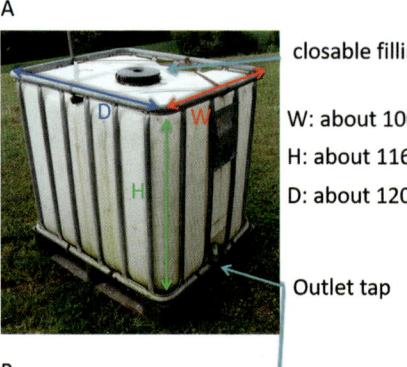

A

closable filling port

W: about 1000 mm
H: about 1160 mm
D: about 1200 mm

Outlet tap

B

D1: 60 mm, DN 50 (S60X6) coarse thread to 2" fine thread
P1: 6 mm

Storz-coupling 2"
(C-pipe, 52 mm)

Fig. 8.9. 1.000 l IBC water tank (A) with adapter set from coarse to fine thread (B).

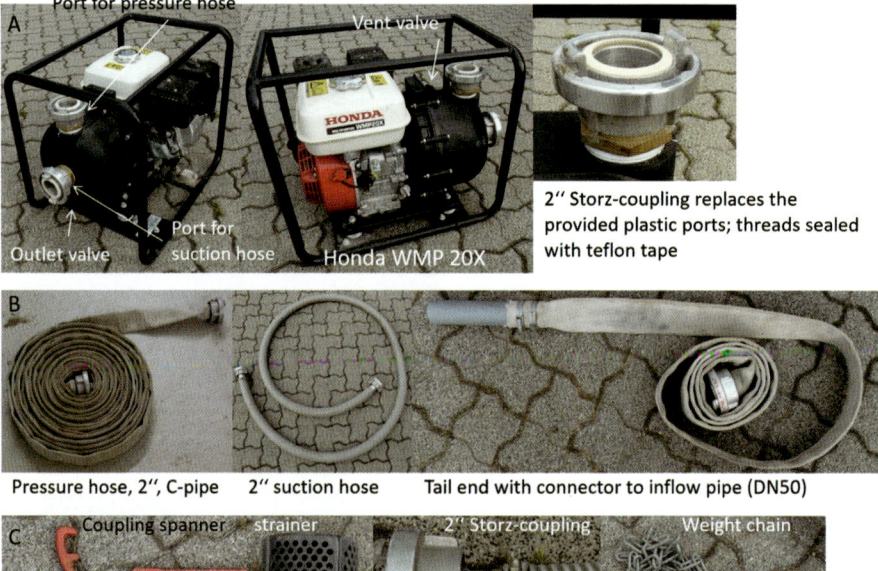

Fig. 8.10. Pump with the different ports (A); different pipes in use (B); important tools and small parts (C).

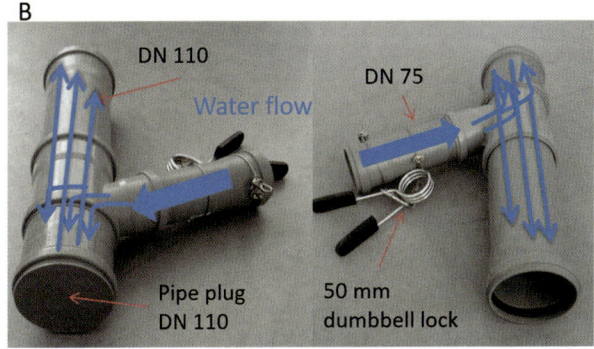

Fig. 8.11. Inflow pipe in use (A); schematic operation procedure (B).

having adapters to connect 2" hoses to them (Fig. 8.9B). To make such connections, suitable spanners are required (Fig. 8.10C).

The water is pumped into the rill directly from the tank. As the pump causes a considerable head at the end of the hose, an implement is required to reduce the pressure and prevent mobilizing material at the starting point of the experiment that is actually unrelated to actual runoff. This implement consists of a pipe, fixed in the soil by means of pipe brackets, onto which a metal spike was welded, which is pushed into the soil and a rubber floor mat held in place at every corner by four tent pegs. The lowermost edge marks the zero point of the rill experiment. Water is moved to that pipe using 2" fire hoses (C-pipe) with a straight DN pipe (bulges must be removed with a metal saw) attached to its end by means of pipe brackets (Fig. 8.10B). The hoses must be arranged in wide arcs and may not have any sharp bends or kinks. The pipe laid in the direction of runoff is a plastic pipe DN 110, 250 mm (Fig. 8.11) and a branching tube DN 75, through which the water enters and after reducing its head flows onto the rubber mat and from there into the rill experiment. The pipe construction is fixed by two pipe clamp welded on spikes. A dumbbell lock is used to attach the fire hose to the DN 75 branch piece. A flow meter may additionally be installed where the hose enters the DN75 branch piece to reduce runoff intensity, where needed. In any event, the setup reduces the head of the water coming from the tank, that goes into the rill.

Tape measure: Along the length of the rill, a tape measure is laid out and fixed with tent pegs. The tape (0) starts at the lower margin of the rubber mat. The tape measure is to follow the course of the rill, its labeling must be well readable, every full meter should carry a distinctly colored additional mark (yellow in Fig. 8.12).

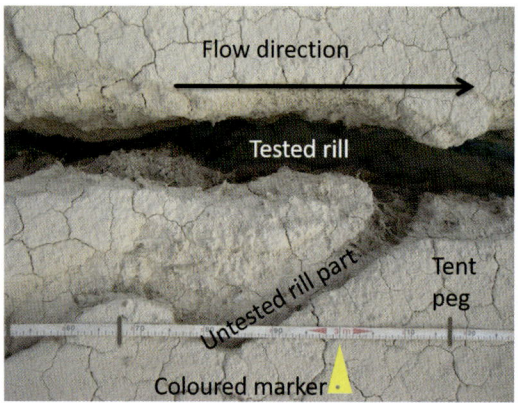

Fig. 8.12. Tape measure with marker for full meter distances.

Runoff measuring station: At the terminus of the rill section to be studied, a runoff measurement plate is installed. It consists of a stainless steel plate of 5 mm thickness to the short sides of which steel of 35 cm length were welded, which are inserted into the soil to keep the plate in place. The plate has a central hole of 12.5 cm diameter with a connector (7.5 cm long) to which a section of DN 125 of 50 cm length is attached. At the end of the pipe section another pipe with an 87° branch is added, which bears a hole of 5 cm diameter above the branch (Fig. 8.13B). The measurement plate is first installed by driving the two lateral spikes into the ground with a plastic or rubber hammer until the lower level of the runoff hole has reached the ground level of the rill.

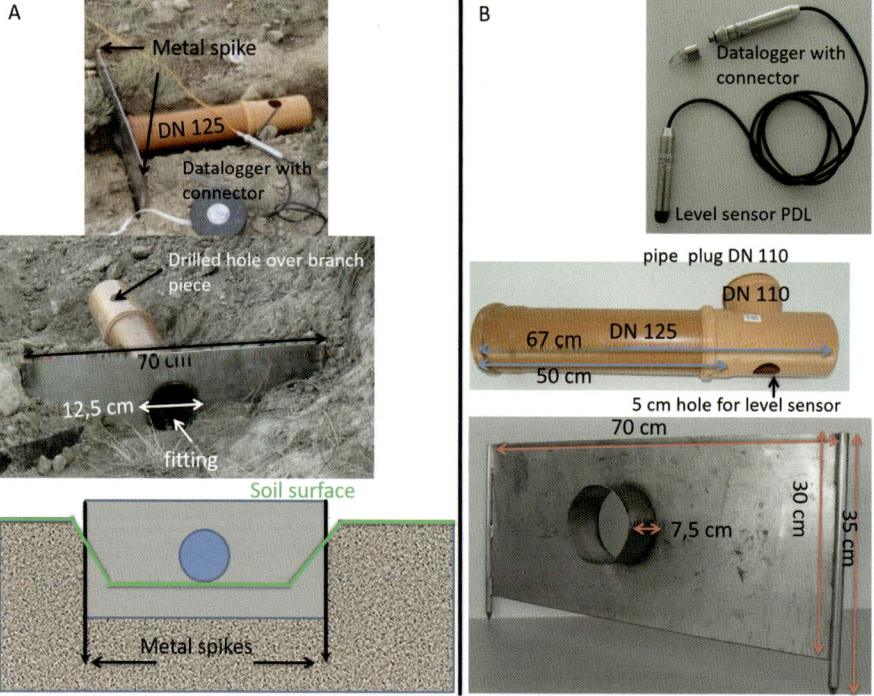

Fig. 8.13. Runoff measuring station: installed into a rill (A); single parts (B).

The housing for measuring the water level is attached at the downward facing side of the plate, which itself is leaning slightly downhill. The measurement plate can by further stabilized with soil. However, care must be taken that no soil material is removed or added uphill from the plate. Below the plate, adding material even to the rill is unproblematic. The DN 125 pipe is now connected to the measurement plated, to that the level housing faces downhill from the plate. To do this, it is often time required to dig into the rill, where the level housing is located. For this reason the level housing (branch) should be as far away from the measuring plate as possible, because digging a hole into the rill could cause water washing out under the measuring plate. At the end of the pipe, a hole is dug which must be large enough to easily house a 10 l bucket. Instead of using a DM 125 pipe, any other pipe can be used, provide it does not limit the actual runoff, i.e. if it its diameter large enough to handle the actual runoff. If the measurement plate is installed at the terminus of a rill, a relief runoff ditch must be dug to prevent water from damming up there. If the rill continues downhill of the plate, make sure that there are no obstacles which impede the runoff that comes off the pipe.

Water level: Into the hole above the branching piece a water level logger is inserted. It automatically and continuously determines the water level by recording the absolute pressure of the water column above it, compensating fluctuations of atmospheric pressure. The branch piece should be filled with water, so that there already is some pressure on the instrument. The water level data should be recorded and logged at 1–2 second intervals. Furthermore, the clocks of the probe and the recording laptop must be properly synchronized and the end of recording should be set, making sure that it does not stop prematurely before the entire experiment has been carried out. By calibrating (generating runoff with known volumes of water and a stopwatch), pressure values obtained from the level probe are converted to runoff volumes.

Measuring locations: In three locations along the rill profile measuring equipment is installed, so called measuring bridges, which carry a centrally attached ultrasonic sensor (e.g., BUS 300 T/UI or USM 30-3000) and also carry a laser distance measuring unit which can be moved along the length of the bridge. The laser distance meter is used to measure the topographic profile across the rill before and after the experiment at the measurement location. The ultrasonic sensor serves to measure the changing water level during the experiment. Additional hydraulic parameters may be computed from these two quantities. The measurement bridges are manufactured from two table bases each. The threads of the furniture bases are hollowed out with a drill and are connected with a threaded rod of 34 cm length (Fig. 8.14A–C). Adjusting the female screws allows to bring the bridge in a level position over the rill. To improve its stability, the bottom of the furniture base is screwed on to a wider plate, which is secured in the soil with tent pegs or large nails.

Additionally, the entire bridge is securely attached to the ground with two steel cables. The cables are to be marked with fluorescent tape to prevent people from stumbling over the steel cables and destruction of the bridge. To prevent the bridge from moving sideways, metal plates, which extend over the top of the furniture bases, are glued onto the lower two furniture bases. A hole of 20 mm diameter (for the ultrasonic sensor) is cut into the surface-parallel, bottom part of the angular profile. During setup of the bridge, this hole should be arranged right above the long axis of the rill bottom, if possible (Fig. 8.14A). A mobile sled, which houses the laser distance meter, mounted on a movable slider is attached to the bridge part vertical to soil surface (Fig. 8.14B). Using threaded rods and female screws allows leveling the setup across the rill. To allow leveling the setup along the length of the rill, the furniture bases should be put onto a level surface. For that purpose terraces must be dug into the slopes. The terraces of either side of the rill should be at about the same level, to allow leveling by adjusting only the nuts. For stability, the distance between the lower furniture bases and the furniture bases of the transverse section should be kept as small as possible. The measuring bridge proper consist of an angle section (steel, perforated 70 x 35 x 1 mm) 70 cm long and 1 mm thick. The surface oriented parallel to

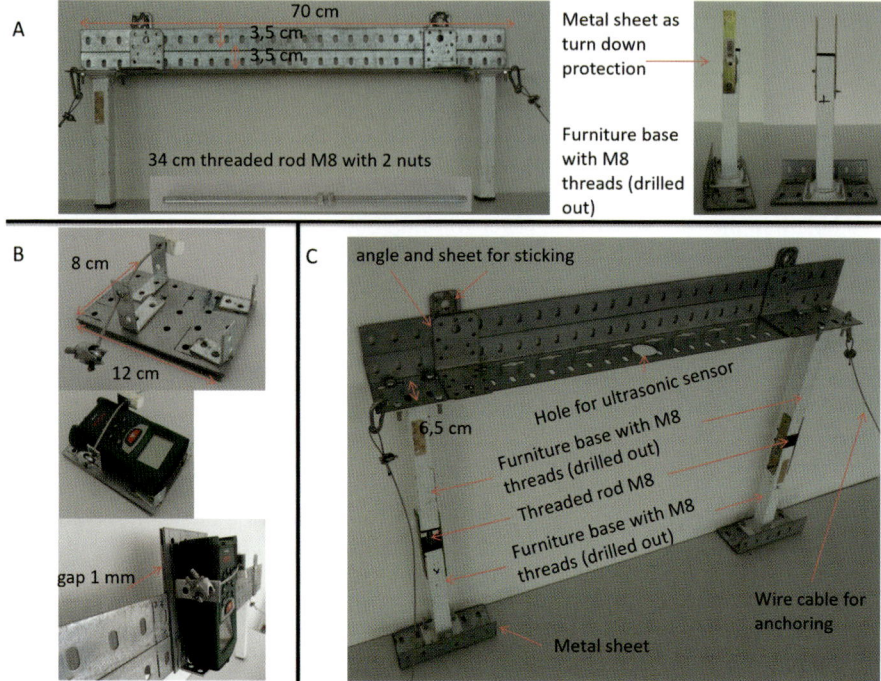

Fig. 8.14. Measuring bridge: bridge parts (A); carriage for the laser rangefinder (B); constructed bridge (C)

the soil surface is 7 cm wide, while the vertical part is 3.5 cm wide, and is widened to 7 cm by attaching a perforated steel plate of 3.5 cm width to it, which receives the mobile sled, with the laser distance meter.

Taking water-sediment samples: Quantification of the amount of eroded material, runoff, and sediment load is achieved by taking water and sediment samples at predetermined times at multiple sampling locations along the rill. To that end, four water and sediment samples are taken at each of the sampling locations. The samples are collected in 1000 cm³ wide neck PET bottles. The timing of sampling is 0:00 (first water arriving at sampling point) and 0:30, 1:30, 2:30 min etc. Water and amount of sediment are determined in the laboratory by weighing after filtering and drying the samples. These data correspond to a spatially and temporally resolved sequence of sediment load. As only partial samples can be taken, the total of eroded material must be calculated from the sediment load and the flux – it cannot be determined by direct measurement. The flux at the individual measuring locations is computed as the product of flow velocity and cross sectional area.

Cross section profiles of the rills at the sampling locations (Fig. 8.15): Using a laser distance meter, cross sectional profiles of the rill are determined before the first experiment, after the first experiment, and after the second experiment (Fig. 8.16). To that end, equidistant markings were made on the angular profile of the measuring bridge, to keep the distance between individual measurements constant (1–2 cm are a good spacing). The values are to be recorded on a protocol prepared for this purpose. For this, make sure that all measurements are always made in the same direction, e.g., from left to right looking uphill, this should also be noted on the protocol. As laser distance meter, for example a Bosch PLR30, may be used, paying attention to the minimum measurement distance of the instrument (which must not be exceeded). Keep in mind that the laser distance meter is attached at a distance of 20–30 cm above the ground, so that

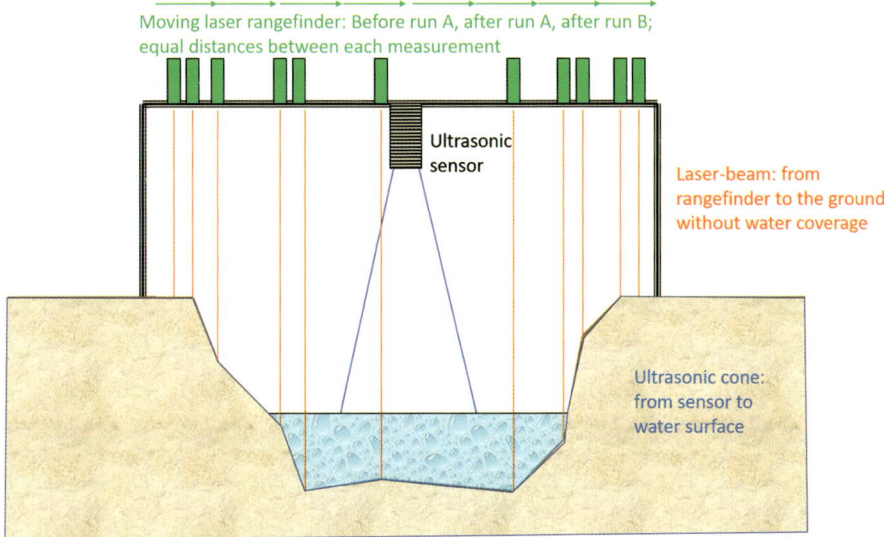

Fig. 8.15. Rill cross section measurement: During execution, the distances between the measurements have to be constant. Note, in this figure only the prominent cross section points are pictured.

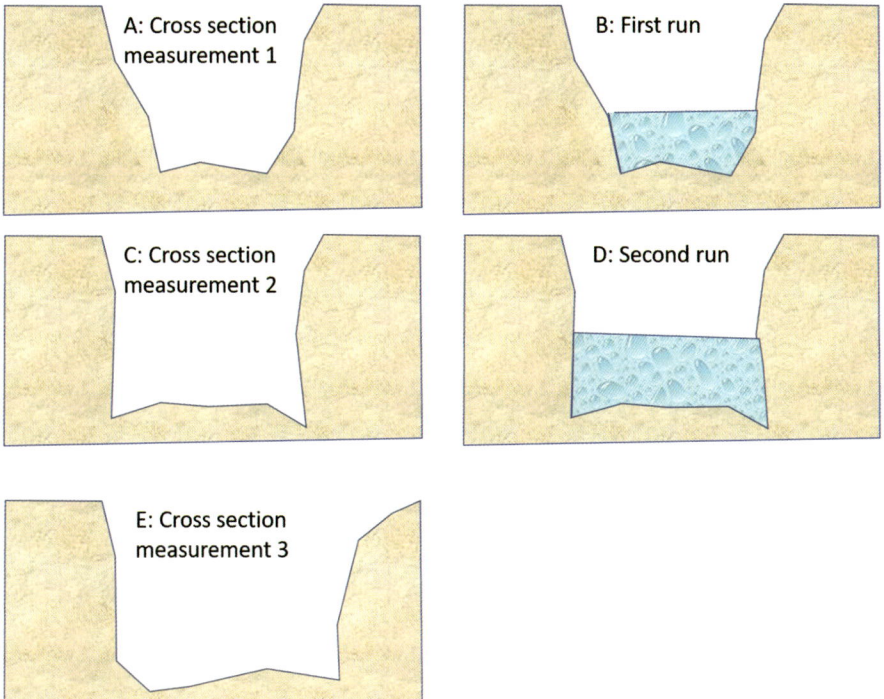

Fig. 8.16. Execution of the cross section measurements: Modifications within the rill are quantifiable.

the minimum measuring distance must be less than the distance between meter and the ground. The sled that holds the distance meter has the shape of a U-profile, the gap is about as thick as the measuring bridge (1 mm). Both plates of the sled are 12 x 8 cm, a metal plate of 3.5 x 8 x 1 mm is glued between the two plates in the upper half. This makes the sled fit perfectly onto the vertical area of the measuring bridge. Four small metal brackets are glued onto the sled, in order to prevent the distance meter from falling off and tilting (Fig. 8.14B). A steel cable between the two upper bracket serves as a "safety belt" for the laser distance meter.

Measuring the water level at the measurement locations: Laser distance meters are unsuitable for measuring water levels, as their laser beam is not reflected by the water surface, but is refracted. Therefore, an ultrasonic water level sensor is used to measure the water level during the experiment. From the rill profile and the water level many important hydraulic parameters are calculated after the experiment (Wirtz et al. 2013a). The ultrasonic water level meter (here a BUS 300 T/UI; Fig. 8.17C shows a similar model of identical shape), which comes in a threaded casing and is mounted in the hole of the measuring bridge and secured with thin nuts on the top and the bottom of the sensor. The sensor comes with an USB adapter, which is used to transfer the data to a computer and supplies the sensor with power. In order to connect multiple sensors to a single computer, we used a Digitus USB Extender "Remote" (5V). The extender receives 5V power from a Yuasa lead accumulator (NP3.2-12-12V, 3.2 Ah; Fig. 8.17C). A suitable voltage converter must be employed to reduce the 12 V supplied by the Pb-battery to 5V required by the Digitus USB-Extender. Here, you should consult an electronics expert in order to prevent damaging the ultrasonic sensor. The Digitus extender remote (probe) is connected to the Digitus ex-

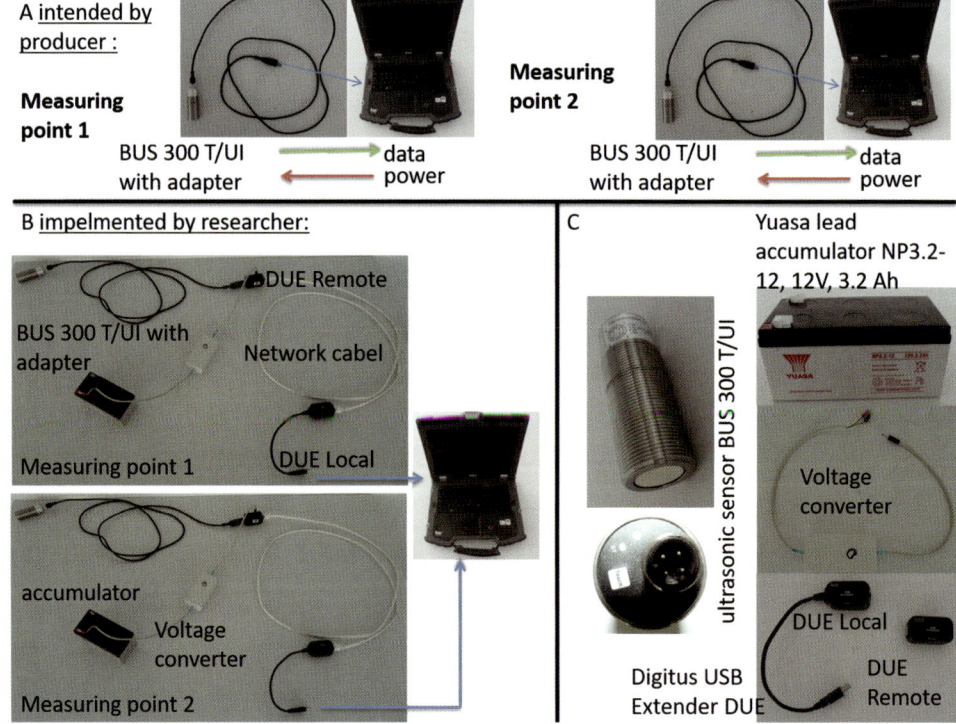

Fig. 8.17. Intended installation of the ultrasonic sensor (A); implemented installation of the supersonic sensor (B); detailed view of the parts (C).

tender Local (computers) using a standard CAT5 network cable, the latter has an USB connector that plugs into the computer. This setup is verified to bridge distances of up to 30 m (Fig. 8.17B). The minimum range of the ultrasonic probe should not be more than 20 cm (measurement range of BUS 300 T/UI is 20–300 mm). Furthermore, the aperture angle of the ultrasonic cone must be checked (11° for the BUS 300 T/UI). It must be ensured that the combination of minimum range and aperture cone allows a measurement within the rill without areas outside the rill.

Measuring the flow velocity of water: Two color tracers (useful are a lighter color (red) and a darker color (blue)) and three stopwatches (with lap capability) are used to measure flow velocities in the rill, which allows measuring three flow velocities: velocity of the water front, velocity of tracer 1 (red), and velocity of tracer 2 (blue). Repeating the tracer measurements multiple times allows documenting changes of the flow velocity during the experiment. For environmental reasons, we recommend to use food colors (e.g., blue E131 and Azorubin E122; Fig 18). Approximately two tablespoons of the food color are put into a container, which is deposited at the starting point of the flow velocity measurement. The dye is not mixed with water but simply poured into the water flowing in the rill.

Fig. 8.18. Azorubin color tracer (A) in action (B).

Number of stopwatches required: At the starting point one stopwatch is required, which record the absolute time of the experiment and the times when the dyes are added to the flowing water (master stopwatch 1). Additionally, three stopwatches with split time capability are required for the "runners" (flow velocities of the water front, tracer 1 and tracer 2). At the location of runoff measurement, two stopwatches are required, one to time the duration of filling of the calibration vessels, the other to document the absolute times of the calibration measurement (master stopwatch 2). A stopwatch is required for every measurement location, to determine the relative times of sampling. These stopwatches are started the moment the water front arrives at that location.

Personnel required: 3 runners to measure flow velocities, 3 samplers at the measuring locations, 1 pump operator (pump operation, master stopwatch 1; announcing the addition of tracers and controlling the length of the experiment) at the top end of the rill and 2 people operating the runoff measuring location (1 x filling the calibration vessels, timing of how long this takes using master stopwatch 2). Doubling of personnel at the measuring locations is helpful but not required.

8.3.3.2. Instructions for setting up the experiment in the field

A rill experiment should always consist of two runs, to take into account the natural soil water content before the rill experiment. If the soil is water saturated prior to the first run, a second run may be dispensable, as there would be no notable difference in soil water content between the beginning of the first run and the beginning of the second one. The effect of initial soil water content on sediment transport is described by Govers et al. (1990) and Govers (1991). Their experiments show that the sediment transport approaches the transport capacity of the rill under dry conditions, while at higher initial soil water content sediment concentrations are lower. Higher initial soil water content reduces the infiltration of water into the soil, which results in a higher volume of water rushing down the rill section. At the same time, this limits the amount of particulate matter delaminated from the soil surface, thus reducing the amount of sediment transported. Here, one must differentiate whether on deals with a soil whose cohesion has been increased by wetting, or a water saturated soil which features low cohesion (reduced attachment of individual grains). Greater soil water content at the soil surface also reduces water repellency of the soil surface, which can be important for runoff dynamics, particularly where soils are normally dry.

The experiment is to prepare as follows:

Step	Action
1	Define starting point and end point of experiment
2	Prepare measureing tape, follow rill course, each meter marked by colored label
3	Install inflow construction; down slope end of the doormat is the 0-m point on the measuring tape
4	Install runoff-measuring construction
5	Install water level sensor, programming if needed
6	Prepare runoff measuring vessel at the runoff measuring point
7	Install measuring bridges; install, check, calibrate ultrasonic sensors; start recording
8	Measure rill cross section at the sampling points
9	Distribute sampling bottles and stopwatches; prepare color tracers
10	Complete admission forms as far as possible
11	Prepare needed admission form at runoff measuring point
12	Briefing and positioning of personal
13	Hook up pump; open tank oap
14	Start experiment

Fig. 8.19. Checklist for test preparation.

8.3.3.3. Points of failure. What can go wrong during an experiment?

When water sediment samples are collected, make sure that all prepared bottles are used. Hereby, it is not important that all bottles are completely filled, but it is important not to miss a sampling interval. Particularly the 30 s interval between sediment sample 1 and sediment sample 2 is very short. To collect these samples, do not press the bottle rim into to soil nor "shove" sediment into the bottle actively. Only material moved by the running water is to be sampled. Small vertical steps in the longitudinal rill profile have proven to be suitable sampling points, as the sampling bottle just needs to be held underneath them. The top opening of the water tank must of course be open before water can be withdrawn from the bottom. Failing that, creates a negative pressure in the tank (vacuum) that may destroy the tank (by collapsing it). The drain pipe at the runoff measurement point should be selected large enough that it does not throttle the real runoff, i.e., greater than the actual runoff.

8.3.3.4. Results

Flow velocity, water levels, the rill cross sectional areas, and the slope angle are the only quantities that can be directly measured in the field. After filtration of the sediment samples in the laboratory, sediment concentrations at different times are determined for the measurement locations. By converting pressure values from the water level meters (using calibration data) total runoff is obtained. For known volumes of water and runoff, the runoff coefficient, that is the ratio of the amount of water applied to the runoff, is obtained. Knowledge of the rill cross-sectional are and water levels may be used to calculate the flow diameter and wetted perimeter. The hydraulic radius may then be computed as the ratio of these quantities.

8.3.3.5. Evaluating and interpreting the runoff experiment

Next to quantities obtained directly from the experiment, it is necessary to calculate numerous derived quantities from these, as they are required to interpret the experiment, in a meaningful way. A review of the relevant formulas for calculating derived quantities is found in (Wirtz et al. 2013b). Further quantities, derived from the mesurements:

Shear stress (Pa): $\tau = \rho * g * R * S$ (8.1)

Unit length shear force (N m^{-1}): $\Gamma = \rho * g * A * S = \tau * W_P$ (8.2.)

Stream power (W m^{-2}): $\omega = \rho * g * R * S * v = \tau * v$ (8.3)

Unit stream power (m s^{-1}): $\omega_U = S * v$ (8.4)

Effective Stream Power (W m^{-1}): $\omega_{eff} = \dfrac{(\tau * v)^{1.5}}{d^{\frac{2}{3}}} = \dfrac{\omega^{1.5}}{d^{\frac{2}{3}}}$ (8.5)

Transport rate (g s^{-1}): $T_R = \text{SSC} * v * A$ (8.6)

Mass of eroded material (g): $A_t = \text{SSC} * v * A * T$ (8.7)

where ρ is the sample raw density (kg m^{-3}), g is the acceleration of gravity (~9.81 m s^{-2}), R is the hydraulic radius (m), A is the flow cross section (m^2), S is the effective slope (= sin(slope)), W_P is the wettet perimeter (m), v is the flow velocity (m s^{-1}), d is the water depth (m), SSC is the concentration of particulate matter (sediment) (g l^{-1}), t is time (s), whereby sample 1 is averaged using the time the water front takes to flow from the start to the sampling location, sample 2 is the average for the following 30 seconds, sample 3 for the following 60 just as for sample 4.

Relevant processes and their identification: A number of parallel processes active in rill erosion can be identified. First, the loose material in the rill is picked up by the water and moved downrill. This can be gleaned from a high sediment concentration, and the relatively slowly moving water front. Depending on the amount of loose material and the runoff effectiveness, material is picked up right at the waterfront. Alternatively this process requires a period of up to some minutes to complete, which is identifiable by high sediment concentrations in a number of water samples. Transport of loose material at and by the water front is called the bulldozer-effect (Regüés et al. 2000, Seeger et al. 2004, Wirtz et al. 2012b), as a large fraction of the transported material is pushed downhill by the waterfront that follows it. Where this is the only process, the sediment concentration in the water samples decreases continuously from the first to the last water sample. In the second run, the same effect is observed, albeit at lower total sediment concentrations, because, as the flow of water subsides, some amount of particulate material remains in the rill. Next to transporting loose material, the water can incise into the sole of the rill by depth erosion, i.e., the rill profile deepens. This process is predominantly controlled by shear stress and would be identifiable by increasing sediment concentrations in individual sampling locations in the course of the experiment. In the second run, the sediment concentration would remain at high values, as the by now completely water-saturated soil is easier to erode.

Furthermore, some effects are observed, which are only indirectly caused by flowing water. The runoff both undercuts the lateral walls of the rill and by head erosion wastes away ledges in the rill bed. This mobilizes comparatively small amounts of sediment. Larger amounts of sediment are subsequently mobilized by gravitational collapse of the so destabilized rill margins and ledges, which are washed in to the rill and transported similarly to the loose material, washed and transported off at the start of the experiment. *Plunge pool dynamics* is another process capable of providing considerable amounts of sediment in a short period of time: the depression below a ledge fills up with water and once the pool is full, the pool runs over, releasing the sediment deposited in the pool over a long period of time very rapidly. Such release events are identifiable by distinctly increased sediment concentrations in the course of sediment samples of a single measuring location.

The ideal patterns of these processes are obfuscated where more than one process are active concurrently, so that direct observation of the processes in the rill during the experiment furnish important hints for later interpretation.

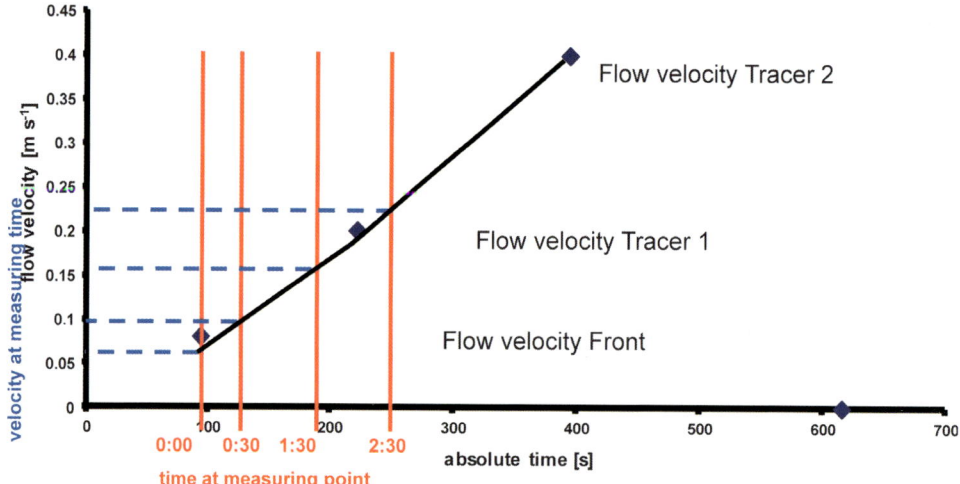

Fig. 8.20. Derivation of flow velocities at the time each individual sample was taken (Wirtz et al. 2013a).

8.3.3.6. Special notes

Please note, that flow velocities measured cannot be directly assigned to individual water/sediment samples directly. In a first step, the relative times of sampling (0:00, 0:30, 1:30, 2:30 min) must be correlated with absolute times of the experiment. Subsequently, it must be checked at what absolute (experiment) times the water front and the tracers arrived at the respective sampling locations and which flow velocities prevailed there. To do this, we assume a linear time flow-velocity relationship to construct a time-flow-velocity diagram in order to assign a flow velocity to any of the sampling times (see Fig. 8.20). Be aware that the assigned flow velocity is not correct because the linear relationship is a severe simplification. But within the frame of the described setup, there is no other option to solve the task joining sediment concentration with a certain flow velocity.

The ultrasonic water level sensors are not waterproof. This means to make sure that particularly the connectors do not get wet and the measuring bridges are protected against falling over (even if rill walls are washed out or collapse). Should it rain during an experiment, the ultrasonic sensors must either be protected against rain or removed.

To measure slope angle, only averages over a meter of longitudinal profile are recorded. Steps and ledges are not considered for the slope angle, but the positions and offset heights should be noted to include them in later interpretations.

8.4. References

Auerswald, K., 1998. Bodenerosion durch Wasser. In: Richter, G. (Ed.), Bodenerosion Analyse und Bilanz eines Umweltproblems. Wiss. Buchges., Darmstadt, pp. 33–42.

Auerswald, K., Fiener, P., Dikau, R., 2009. Rates of sheet and rill erosion in Germany – A meta-analysis. Geomorphology 111: 182–193. doi: 10.1016/j.geomorph.2009.04.018

Bowyer-Bower, T.A.S., Burt, T.P., 1989. Rainfall simulators for investigating soil response to rainfall. Soil Technology 2: 1–16. doi: 10.1016/S0933-3630(89)80002-9

Brunton, D.A., Bryan, R.B., 2000. Rill network development and sediment budgets. Earth Surf. Process. Landforms 25: 783–800. doi: 10.1002/1096-9837(200007)25:7<783::AID-ESP106>3.0.CO;2-W

Bryan, R.B., 2000. Soil erodibility and processes of water erosion on hillslope. Geomorphology 32: 385–415. doi: 10.1016/S0169-555X(99)00105-1

Bryan, R.B., Poesen, J., 1989. Laboratory experiments on the influence of slope length on runoff, percolation and rill development. Earth Surf. Process. Landforms 14: 211–231. doi: 10.1002/esp.3290140304

Casalí, J., Loizu, J., Campo, M.A., De Santisteban, L.M., Álvarez-Mozos, J., 2006. Accuracy of methods for field assessment of rill and ephemeral gully erosion. Catena 67: 128–138. doi: 10.1016/j.catena.2006.03.005

Clarke, M.A., Walsh, R.P.D., 2007. A portable rainfall simulator for field assessment of splash and slopewash in remote locations. Earth Surface Processes and Landforms 32: 2052–2069. doi: 10.1002/esp.1526

de Lima, J.L.M.P., 1990. The effect of oblique rain on inclined surfaces: A nomograph for the rain-gauge correction factor. Journal of Hydrology 115: 407–412. doi: 10.1016/0022-1694(90)90218-M

de Lima, J.L.M.P., 1989. Raindrop splash anisotropy: Slope, wind and overland flow velocity effects. Soil Technology 2: 71–78. doi: 10.1016/S0933-3630(89)80008-X

Dunne, T., Malmon, D.V., Mudd, S.M., 2010. A rain splash transport equation assimilating field and laboratory measurements. J. Geophys. Res. 115: F01001. doi: 10.1029/2009JF001302

Emde, K., Szöcs, A., 2000. Geoökologische Arbeitsmethoden II. Johannes Gutenberg Universität, Mainz.

Fister, W., Iserloh, T., Ries, J.B., Schmidt, R.-G., 2012. A portable wind and rainfall simulator for in situ soil erosion measurements. Catena 91: 72–84. doi: 10.1016/j.catena.2011.03.002

Furbish, D.J., Hamner, K.K., Schmeeckle, M., Borosund, M.N., Mudd, S.M., 2007. Rain splash of dry sand revealed by high-speed imaging and sticky paper splash targets. Journal of Geophysical Research 112, F01001. doi: 10.1029/2006JF000498

García-Ruiz, J.M., Beguería, S., Nadal-Romero, E., González-Hidalgo, J.C., Lana-Renault, N., Sanjuán,

Y., 2015. A meta-analysis of soil erosion rates across the world. Geomorphology 239: 160–173. doi: 10.1016/j.geomorph.2015.03.008

Gilley, J., Kottwitz, E., Simanton, J., 1990. Hydraulic Characteristics of Rills. Biological Systems Engineering: Papers and Publications.

Giménez, R., Govers, G., 2002. Flow Detachment by Concentrated Flow on Smooth and Irregular Beds. Soil Science Society of America Journal 66: 1475–1483. doi: 10.2136/sssaj2002.1475

Govers, G., 1991. Time-dependency of runoff velocity and erosion the effect of the initial soil moisture profile. Earth Surf. Process. Landforms 16: 713–729. doi: 10.1002/esp.3290160805

Govers, G., Everaert, W., Poesen, J., Rauws, G., De Ploey, J., Lautridou, J.P., 1990. A long flume study of the dynamic factors affecting the resistance of a loamy soil to concentrated flow erosion. Earth Surface Processes and Landforms 15: 313–328. doi: 10.1002/esp.3290150403

Govers, G., Giménez, R., Van Oost, K., 2007. Rill erosion: Exploring the relationship between experiments, modelling and field observations. Earth-Science Reviews 84: 87–102. doi: 10.1016/j.earscirev.2007.06.001

Huang, C., Laflen, J.M., Bradford, J.M., 1996. Evaluation of the Detachment-Transport Coupling Concept in the WEPP Rill Erosion Equation. Soil Science Society of America Journal 60: 734–739. doi: 10.2136/sssaj1996.03615995006000030008x

Iserloh, T., 2013. Niederschlagssimulationen mit kleinen mobilen Beregnungsanlagen – Tropfenerzeugung, Regnervergleich, windbeeinflusster Niederschlag. (Dissertation). Universität Trier, Trier.

Iserloh, T., Fister, W., Marzen, M., Seeger, M., Kuhn, N.J., Ries, J.B., 2013a. The role of wind-driven rain for soil erosion – an experimental approach. Zeitschrift für Geomorphologie, Supplementary Issues 57: 193–201. doi: 10.1127/0372-8854/2012/S-00118

Iserloh, T., Ries, J.B., Arnáez, J., Boix-Fayos, C., Butzen, V., Cerdà, A., Echeverría, M.T., Fernández-Gálvez, J., Fister, W., Geißler, C., Gómez, J.A., Gómez-Macpherson, H., Kuhn, N.J., Lázaro, R., León, F.J., Martínez-Mena, M., Martínez-Murillo, J.F., Marzen, M., Mingorance, M.D., Ortigosa, L., Peters, P., Regüés, D., Ruiz-Sinoga, J.D., Scholten, T., Seeger, M., Solé-Benet, A., Wengel, R., Wirtz, S., 2013b. European small portable rainfall simulators: A comparison of rainfall characteristics. Catena 110: 100–112. doi: 10.1016/j.catena.2013.05.013

Iserloh, T., Ries, J.B., Cerdà, A., Echeverría, M.T., Fister, W., Geißler, C., Kuhn, N.J., León, F.J., Peters, P., Schindewolf, M., Schmidt, J., Scholten, T., Seeger, M., 2013c. Comparative measurements with seven rainfall simulators on uniform bare fallow land. Zeitschrift für Geomorphologie, Supplementary Issues 57: 11–26. doi: 10.1127/0372-8854/2012/S-00085

Kinnell, P.I.A., 2005. Raindrop-impact-induced erosion processes and prediction: a review. Hydrological Processes 19: 2815–2844.

Knapen, A., Poesen, J., Govers, G., Gyssels, G., Nachtergaele, J., 2007. Resistance of soils to concentrated flow erosion: A review. Earth-Science Reviews 80: 75–109. doi: 10.1016/j.earscirev.2006.08.001

Kuhn, N.J., Greenwood, P., Fister, W., 2014. Chapter 5.1 - Use of Field Experiments in Soil Erosion Research. In: Thornbush, M.J., Allen, C.D., Fitzpatrick, F.A. (Eds.), Geomorphological Fieldwork. Elsevier, Amsterdam, pp. 175–200.

Le Bissonnais, Y., 1996. Aggregate stability and assessment of soil crustability and erodibility: I. Theory and methodology. European Journal of Soil Science 47: 425–437. doi: 10.1111/j.1365-2389.1996.tb01843.x

Legout, C., Leguédois, S., Le Bissonnais, Y., Malam Issa, O., 2005. Splash distance and size distributions for various soils. Geoderma 124: 279–292. doi: 10.1016/j.geoderma.2004.05.006

Leguédois, S., Planchon, O., Legout, C., Bissonnais, Y.L., 2005. Splash projection distance for aggregated soils: Theory and experiment. Soil Science Society of America Journal 69: 30–37.

Mancilla, G.A., Chen, S., McCool, D.K., 2005. Rill density prediction and flow velocity distributions on agricultural areas in the Pacific Northwest. Soil and Tillage Research 84: 54–66. doi: 10.1016/j.still.2004.10.002

Martínez-Murillo, J.F., Nadal-Romero, E., Regüés, D., Cerdà, A., Poesen, J., 2013. Soil erosion and hydrology of the western Mediterranean badlands throughout rainfall simulation experiments: A review. CATENA, Updating Badlands Research 106: 101–112. doi: 10.1016/j.catena.2012.06.001

Marzen, M., Iserloh, T., Casper, M.C., Ries, J.B., 2015. Quantification of particle detachment by rain splash and wind-driven rain splash. Catena 127: 135–141. doi: 10.1016/j.catena.2014.12.023

Marzen, M., Iserloh, T., de Lima, J.L.M.P., Ries, J.B., 2016. The effect of rain, wind-driven rain and wind on particle transport under controlled laboratory conditions. Catena 145: 47–55. doi: 10.1016/j.catena.2016.05.018

Marzen, M. & Iserloh, T. & De Lima, J.L.M.P. & Fister, W. & Ries, J.B. (2017): Impact of severe rain storms on soil erosion: Experimental evaluation of wind-driven rain and its implications for natural hazard management. Science of the Total Environment 590–591: 502–513. doi: 10.1016/j.scitotenv.2017.02.190.

Merz, W., Bryan, R.B., 1993. Critical conditions for rill initiation on sandy loam Brunisols: laboratory and field experiments in southern Ontario, Canada. Geoderma 57: 357–385. doi: 10.1016/0016-7061(93)90050-U

Miehlich, G., 2003. Die Bekämpfung der Bodendegradation. Eine weltweite Herausforderung. Petermanns Geographische Mitteilungen 147: 6–13.

Morgan, R.P.C., 2005. Soil erosion and conservation. 3rd ed., Wiley-Blackwell, Maiden, Oxford, Carlton.

Oldeman, L.R., Hakkeling, R.T.A., Sombroek, W.G., 1991. World Map of the Status of Human-induced Soil Degradation (Revised ed.). Three maps and explanatory note. ISRIC, Wageningen and UNEP, Nairobi.

Poesen, J., 1987. Transport of rock fragments by rill flow – a field study. Catena Supplement 8: 35–54.

Regüés, D., Balasch, J.C., Castelltort, X., Soler, M., 2000. Relación entre las tendencias temporales de producción y transporte de sedimentos y las condiciones climáticas en una pequeña cuenca de la montaña mediterránea (Vallcebre, Pirineos orientales). Cuadernos de Investigación Geográfica 41–66.

Richter, G., 1998. Bodenerosion: Analyse und Bilanz eines Umweltproblems. Wissenschaftliche Buchgesellschaft.

Richter, G., 1983. Bodenerosionsmessung und ihre geoökologische Auswertung. In: Leser, H. (Ed.), Veröffentlichungen des 8. Basler Geomethodischen Colloquiums. Bodenerosion als Methodisch Geoökologisches Problem, Geomethodica. Basel, pp. 23–50.

Ries, J.B., 2011. Bodenerosion. In: Gebhardt, H., Glaser, R., Radtke, U., Reuber, P. (Eds.), Geographie, Physische Geographie und Humangeographie. Spektrum, Heidelberg, pp. 506–515.

Ries, J.B., 2010. Methodologies for soil erosion and land degradation assessment in mediterranean-type ecosystems. Land Degradation & Development 21: 171–187. doi: 10.1002/ldr.943

Ries, J.B., 2001. Geomorphodynamik und Landdegradation zwischen Ebrobecken und Pyrenäen (Habilitation). Universität Frankfurt a. M.

Ries, J.B., Iserloh, T., Seeger, M., Gabriels, D., 2013. Rainfall simulations – constraints, needs and challenges for a future use in soil erosion research. Zeitschrift für Geomorphologie, Supplementary Issues 57: 1–10. doi: 10.1127/0372-8854/2013/S-00130

Ries, J.B., Marzen, M., Iserloh, T., Fister, W., 2014. Soil erosion in Mediterranean landscapes - Experimental investigation on crusted surfaces by means of the Portable Wind and Rainfall Simulator. Journal of Arid Environments 100–101: 42–51. doi: 10.1016/j.jaridenv.2013.10.006

Riezebos, H.T., Epema, G.F., 1985. Drop shape and erosivity part II: Splash detachment, transport and erosivity indices. Earth Surface Processes and Landforms 10, 69–74. doi: 10.1002/esp.3290100109

Roth, C.H., 1995. Physikalische Ursachen der Wassererosion. In: Blume, H.P., Felix-Henningsen, P., Fischer, W.R., Frede, H.-G., Guggenberger, G., Horn, R., Stahr, K. (Eds.), Handbuch Der Bodenkunde. Wiley-VCH, Weinheim, p. 3584.

Saleh, A., 1993. Soil roughness measurement: Chain method. Journal of Soil and Water Conservation 48: 527–529.

Schmidt, J., Werner, M. v., Schindewolf, M., 2017. Wind effects on soil erosion by water – A sensitivity analysis using model simulations on catchment scale. Catena 148: 168–175. doi: 10.1016/j.catena.2016.03.035

Schwertmann, U., Vogl, W., Kainz, M., 1990. Bodenerosion durch Wasser: Vorhersage des Abtrags und Bewertung von Gegenmaßnahmen. 2nd ed., Ulmer, Stuttgart.

Seeger, M., Errea, M.-P., Beguería, S., Arnáez, J., Martí, C., García-Ruiz, J.M., 2004. Catchment soil moisture and rainfall characteristics as determinant factors for discharge/suspended sediment hysteretic loops in a small headwater catchment in the Spanish pyrenees. Journal of Hydrology 288: 299–311. doi: 10.1016/j.jhydrol.2003.10.012

Stroosnijder, L., 2005. Measurement of erosion: Is it possible? Catena 64: 162–173. doi: 10.1016/j.catena.2005.08.004

van Dijk, A.I.J.M., Bruijnzeel, L.A., Rosewell, C.J., 2002a. Rainfall intensity–kinetic energy relationships: a critical literature appraisal. Journal of Hydrology 261: 1–23. doi: 10.1016/S0022-1694(02)00020-3 van Dijk, A.I.J.M., Meesters, A.G.C.A., Bruijnzeel, L.A., 2002b. Exponential Distribution Theory and the Interpretation of Splash Detachment and Transport Experiments. Soil Science Society of America Journal 66: 1466. doi: 10.2136/sssaj2002.1466

Vannoppen, W., Vanmaercke, M., De Baets, S., Poesen, J., 2015. A review of the mechanical effects of plant roots on concentrated flow erosion rates. Earth-Science Reviews 150: 666–678. doi: 10.1016/j.earscirev.2015.08.011

Wirtz, S., Iserloh, T., Rock, G., Hansen, R., Marzen, M., Seeger, M., Betz, S., Remke, A., Wengel, R., Butzen, V., Ries, J.B., 2012a. Soil Erosion on Abandoned Land in Andalusia: A Comparison of Interrill- and Rill Erosion Rates. ISRN Soil Science 2012: 1–16. doi: 10.5402/2012/730870

Wirtz, S., Seeger, M., Remke, A., Wengel, R., Wagner, J.-F., Ries, J.B., 2013a. Do deterministic sediment detachment and transport equations adequately represent the process-interactions in eroding rills? An experimental field study. Catena 101: 61–78. doi: 10.1016/j.catena.2012.10.003

Wirtz, S., Seeger, M., Ries, J.B., 2012b. Field experiments for understanding and quantification of rill erosion processes. Catena 91: 21–34. doi: 10.1016/j.catena.2010.12.002

Wirtz, S., Seeger, M., Zell, A., Wagner, C., Wagner, J.-F., Ries, J.B., 2013b. Applicability of Different Hydraulic Parameters to Describe Soil Detachment in Eroding Rills. PLoS ONE 8: e64861. doi: 10.1371/journal.pone.0064861.

9. Penetration resistance

Heinz Peter Schrey

9.1. Introduction

The penetration resistance of a soil is the reactive force of a soil against imprinting objects, like roots, posts, pegs, or hoof-, wheel-, track imprints. This applies to strip footings, raindrops, or projectiles as well.

At the microscopic level, penetration resistance results from resistance of individual soil particles (mineral grains in coarse or fine sandy soils or aggregates in soils with a developed structure) against being dislocated or from the actual dislocation of these particles, which is a function of water content and bulk density.

The resistance of particles against being dislocated increases:
- where mineral grains and hard aggregates are angular rather than round
- where grain contacts are cemented by smaller particles. This is a result of soil development caused by receding soil water menisci in the drying phases of wetting-drying cycles, which move fine grained material into the apices of the inter-grain spaces, where the grains abut to each other
- where grain contacts are subject to burial pressure at greater depths
- when the contact number of the grains or aggregates increases. This happens when the grain size distribution curve of a soil resembles a Fuller-curve (known from concrete with optimal packing)
- when loose aggregates or aggregate packs are compacted by aggregate destruction
- when poorly structured soils are compacted by loading or initial dewatering

Penetration resistance is thus a cumulative property and depends, at a macroscopic level, on numerous soil properties, particularly on:
- grain size (texture): The respective fractions of sand, silt, and clay do not control penetration resistance in any particular direction. Greater penetration resistance is observed in soils of high shear strength and those that have a high fraction of medium sized and small pores. Furthermore water conductivity, which itself is controlled by aggregation, influences penetration resistance
- bulk density: Higher bulk density in soils with equivalent grain size distributions results in greater penetration resistance. For example, soils at the plow depth in plowed fields or cultivated plots
- solidification: Initially loose mineral particles may be cemented by the formation of Fe-crust, hardpans, and calcretes, which increase penetration resistance at otherwise low bulk densities and high water conductivities
- soil water content (soil water tension): High soil water content at equal bulk density results in low penetration resistance

- caution: This may lead to misinterpretation on a soil which is superficially dried, but with increasing depth still moist. This will yield higher penetrometer readings near surface and lower or equal with increasing depth due to the higher water content
- content of organic matter: Penetration resistance *decreases* as the content of organic matter *increases*
- fraction of gravel, cobble, or boulder in the soil: A greater fraction, at equal or slightly increased bulk density, *increases* penetration resistance because displacing gravel or cobble requires greater force than just displacing small particles

The penetration resistance of a soil is subject to continuous development, both as the soil itself develops and numerous external conditions (precipitation, drought) change, without one being able to attribute the changes in penetration resistance to individual impacts. Thus, penetration resistance readings are generally *not normally distributed*, but instead vary, often erratically, with location and depth.

9.1.1. Basics of measuring penetration resistances

Penetration resistance is measured as the force required for a rod to enter the soil. If the measured force is related to the cross-sectional area of this rod, a pressure is obtained (see Table 9.1).

Table 9.1. Maximum pressures a person weighing 75 kg exerts on the cross-sectional area of different penetrometer tips (60° angle variety).

Height of sounding tip (cm)	1.0	1.4	1.8	2.2
Cross sectional area of tip (cm²)	1.0	2.0	3.3	5.0
Penetrometer rod diameter (cm) (to minimize shaft friction)	0.8	1.0	1.5	1.5
Maximum pressure at tip [kPa]	7500	3750	2273	1500
Maximum pressure [MPa]	7.50	3.75	2.27	1.50

9.1.2. Methods for measuring penetration resistance

This chapter introduces hand-operated instruments only, which differ, however, in the way force (penetration resistance) is measured:

- manual pressure/force sensors
- number of drops of a known weight
- registration of pressure/force with a spring or a manometer

The methods are described in section 2. The field methods to measure penetration resistance up to 1 m depth are discussed considering the following criteria:

- objectivity of the measurement
- sensitivity to the expected soil material
- workability

9.2. Selection of the most suitable method

To determine the penetration resistance of soils, three types of hand-operated devices can be distinguished, which differ in the way they register force or pressure and in the way they are used in the field.

9.2.1 Manual measurement of pressure or force of penetration

The easiest estimate of penetration resistance of a soil is obtained by manually pushing a large nail, spatula, or pocket knife into the soil, particularly into the faces of a soil profile. Alternatively, a soil penetration probe may be used at the soil surface. Such a probe normally consists of a stainless steel rod with a sharpened tip on one end and a ball shaped grip on the other. Characteristic for these methods are their short reach and their qualitative results, because the perceived force required to make the probe enter the soil cannot be quantified. Despite these shortcomings, this method is capable of producing a first assessment of penetration resistance at a profile wall or the soil surface, provided that the measurements are conducted in a short period of time. Intensity and dynamics of the entrance of the probe, applied force, and hence the estimate of penetration resistance are, however, quite subjective and make a comparison to similar measurements possibly made by other operators and at different times and/or places very difficult or impossible. These methods are best suited for used by experienced persons, within a short period of time at a single location, as they require little effort and are easily carried out.

9.2.2. Number of impacts of a falling weight

The Lightweight penetrometer according to ISO 22476-2:2005, also called Dynamic Probing, allows applying a reproducible amount of force onto the probe. It normally consists of a penetrating rod, with a standardized, conic tip along which a weight with a central hole falls onto an anvil attached to the lower part of the rod. Lifting the weight and dropping it from a predefined height onto the anvil then drives the tip into the soil. Standardized probes have a weight of 10 kg, a drop height of 50 cm and a cross section of the penetrating tip of 5 cm^2 (The German industrial norm DIN 4094-3 abbreviates them as DPL-5, previously also LRS 5; also known as "Künzelstab" (Künzel-rod) or with a tip of 10 cm^2 cross section: DPL, previously "LRS 10").

The number of drops of the weight necessary to drive the rod into the soil for a given depth of penetration or increment is recorded, e.g., the number of drops required to make the probe tip enter the soil by 10 cm, or alternatively, the depth of penetration is recorded for a given number of drops. The results of such experiments are generally represented in a step-diagram, which plots the number of drops against the depth of penetration. Defining weight, drop height, and shape and cross section of the probe's tip ascertain that intensity and dynamics of applied force are reproducible. If the penetration depth is determined exactly, such measurements are objective and reproducible.

Typically, Lightweight penetrometers are used for:

- locating the depth of the boundaries of soil layers (subsurface studies for bridges, buildings, road construction projects)
- verifying compaction after compaction measures and piling of soil
- determining the structure of cohesive soils
- obtaining soil physical parameters to assess the maximum load-bearing capacity of soils and for performing stability calculations, particularly because of the good reproducibility of the method.

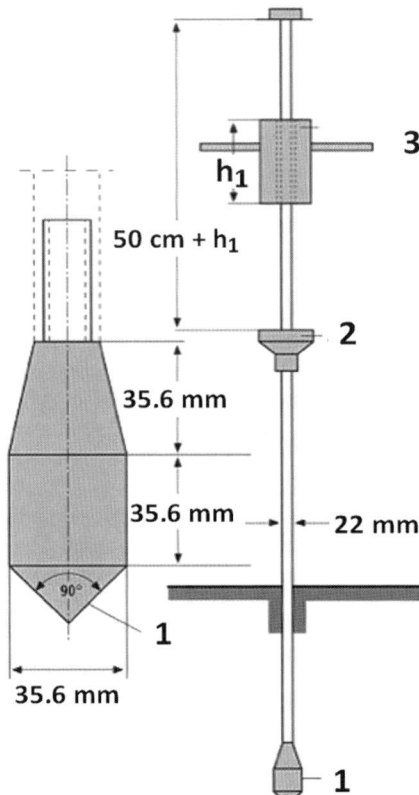

Fig. 9.1. Lightweight dynamic penetrometer "DPL 10" according to DIN EN ISO 22476-2:2012-03 with a falling height of 50 cm. 1 – sounding tip with an opening angle of 90°, 2 – anvil, 3 – weight (hammer, rammer) of 10 kg.

- determining the ability of ramming or compressing the soil, due the comparability of percussion probe results for driving posts into the soil.
- probing not too firm soils to depths of up to 12 m

These fields of application are interested in determining larger variations of soil parameters than are typical for the top 2 m of soil profiles, which are an ecological reactor, host plants, and are relevant for the water and air budget of the soil. In addition, with the advent of industrial norm DIN ISO 22476-2 the smaller tip cross-section of 5 cm² ("Künzelstab") was abandoned in favor of the 10 cm² tip. While the smaller tip required up to 30 drops to advance by 10 cm in compacted soils, the larger tip has increased that number, and by this, the vibrations due to ramming. This increases the probability of affecting the structure of aggregated soils, which is generally high, where penetration probe tips advance jerkily. Furthermore, the larger tip displaces a larger soil volume and so increases the probability of creating temporary excess pore water pressures.

In terms of operability, the percussion probe is operated manually, but transporting its weight (10 kg) and the investment of human effort and time for the required number of weight drops per location considerably limits its applicability over large areas to be studied.

The percussion penetrometer therefore is a well known, widely used instrument, the results of which are comparable to a large number of measurements, delivering data to be used in soil mechanical assessments to derive other soil parameters. It is not suited to identify differences in soil structure or differences in bulk density, which control the soil's water and air budget and are thus ecologically relevant to plant growth. However, it may be used to determine the standard plowing depth (plough pan).

9.2.3. Force measured by a spring or a manometer

Instruments measuring the penetration force by a spring or a manometer, are generally called penetrometers, discounting their scope or field of application for fieldwork they are available in a number of designs with great differences in the way they register and present the measured values.

Schematically, a penetrometer consists of:

- a stainless steel rod, 90 to 110 cm in length, which can be extended by attaching similar rods to one end,
- which carries a sounding tip (cone) with a cross section greater than that of the rod and its extensions. The cone at the tip may have a range of angles, most commonly, however, 60°. Note, that only measurements made with the same tips (of the same size, geometry, angle) are comparable.
- the penetrometer features a force meter between the handles or below them.

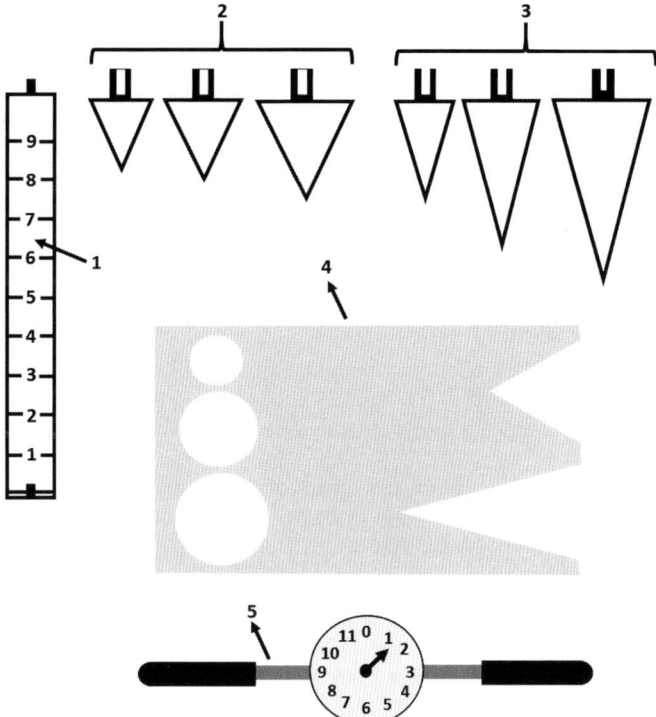

Fig. 9.2. Schematic of a (hand) penetrometer according to DIN 19662. 1 – stainless steel rod, 90 to 110 cm in length; 2 – sounding tips (cone) with different cross sectionsand an opening angle of 60°; 3 – sounding tips (cone) with different cross sections and an opening angle of 30°; 4 – proof sheet with triangles of 30° and 60° and with holes of 97% of the primary sounding tipcross sections; 5 – force meter pointer, maximum pointer and rubber coated handles.

The force meter either is spring based or based on a manometer with

- a drag indicator indicating the maximum force – this instrument is usually called penetrometer or manual penetrometer or
- a recording device, which is called a penetrograph, or alternatively
- a digital device, which is called digital penetrograph or penetrologger, some of which are also equipped with a GPS-receiver or a soil moisture probing tip.

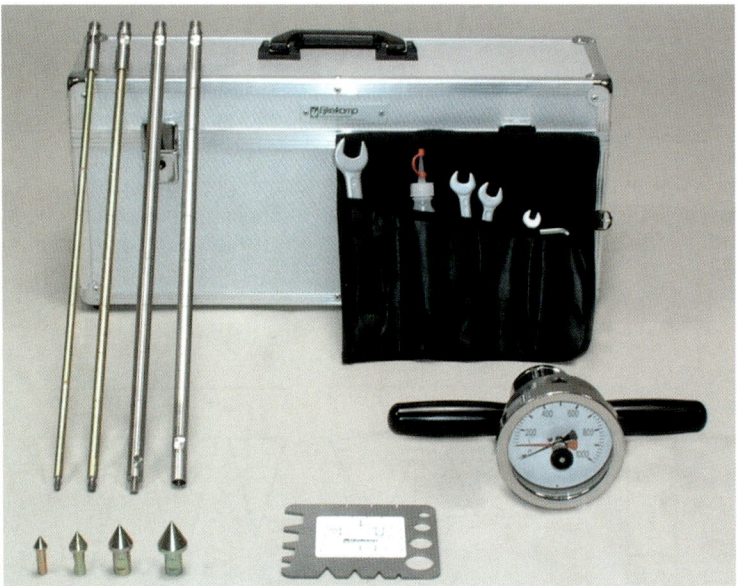

Fig. 9.3. (Manual) penetrometer.

The rod of some smaller models has no separate conical tip but terminates straight in the recording unit, which is attached to the handles used to press the rod into the soil. Such models are called mini-penetrometers and are well suited for (horizontal) measurements of soil profiles in ditches.

Penetrometers measure penetration resistance of field soils up to approximately 1 m depth as a pressure in the range of 0 to 5 Mpa (1 MPa = 0.1 kN cm^{-2} ≡ 10.1972 kg cm^{-2}). Typical depth resolution of the, generally continuous, measurement procedure is 10 cm. With extending rods, measurements down to 2 m depth are possible. Depth resolutions in the cm-range can be achieved by using penetrographs or petrologgers. Measurement is carried out:

- in a single selected spot, or,
- over a (small) area in randomly selected spots,
- in equidistant spots along a transect,
- across a (larger) area in individual points of a grid, with equidistant points in perpendicular directions

If the cross-sectional area of the tip is properly selected to match the problem at hand, (manual) penetrometers are quite well suited for indicating inhomogeneities of the soil properties listed above.

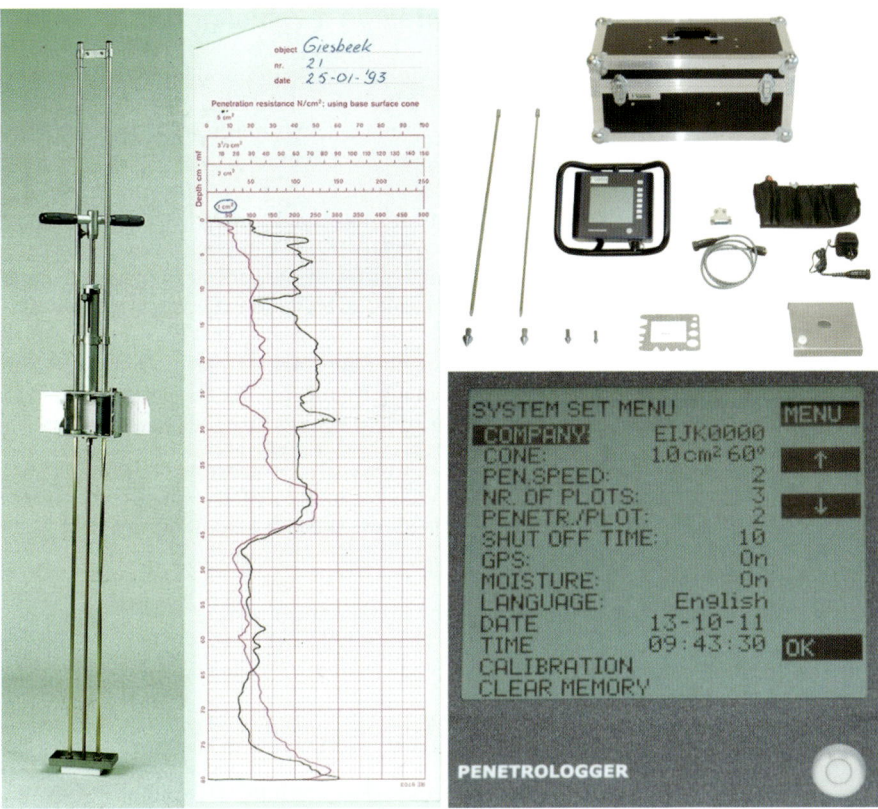

Fig. 9.4. Penetrograph (left) with recording strip (center), and Penetrologger (right top) with screenshot of operating screen (right bottom).

For applications and series of measurements, with typical depth resolutions and about 50 or 100 individual spots, it is recommended to use a manual penetrometer. But if one has access to, and is by frequent use experienced in operating, a GPS-equipped penetrologger, its application may be more effective.

Penetrologgers are advantageous because they feature higher depth resolution and a person (to write down measured values) is not required for their operation. On the other hand, some describe the operation of penetrologgers as cumbersome, once faulty readings are produced or depths exceeding 1.5 m are to be probed.

A GPS-receiver built into the penetrologger is advantageous, if location and instrument allow locating the measuring spot to 0.5 m or at least better than 1 m. Soil water content sensors at the penetrologger tip are less useful, because soil water content measurements are usually made at a constant depth and not continuously. This precludes the use of these data with corresponding penetration resistances measured.

Where the soil water content is unknown, but considered a critical quantity, it is recommended to use soil water content sensors or take soil samples from the disired depth to determine the soil water content at the laboratory after the penetrologger measurements. Before a penetrologger can be used, its supply with electrical power must be ascertained, by (charged!) batteries and some stand-by ones.

9.2.4. Mobile field penetrometer mounted on a vehicle

Over the last years, different institutes developed mobile electronic field penetrometers especially for the finely resolved determination of the penetration resistance, which is mounted on a vehicle with a traverse axle system. In some cases, the vehicle is raised with two supporting legs during the measurement of the penetration resistance by a hydraulic system to avoid influences by the spring shock absorbers (e.g., the system developed by the Humboldt University of Berlin, Germany). In the developed system, the software for the electronic field penetrometer records the data, displays it graphically, stores it with the metadata and can be used by a connected laptop within the framework of a measurement campaign for the control of the experiment and for visualization on the spot. It is thus possible to measure even large experimental areas in a short time and with little physical effort.

Reproducibility of the measurements is the main advantage of the mobile electronic field penetrometer. Automated driving of the penetrometer by means of a traversing system and the increased stability due to the hydraulically mounted legs standardize the measurements:

- the penetration speed is always constant; this also applies to soils with very different strengths (dried out hard or soft plastic) or with abrupt changes in the structure (aggregate, plate structure, biogenous large pores, or passages). This allows a depth resolution in centimeter range to be achieved
- the measurement is always perpendicular
- there is no person-dependent component of the measurement by size, force, or fatigue of the measuring person
- storing the acquired data in CSV format (or other formats) enables their use with other software

Nevertheless, there are some possible extensions, very worthwhile to be made:

- the use of a high-resolution GPS, whose position data are integrated into the software, can simplify the spatial representation of the measurements and allow further evaluations
- the expansion of the measurement depth, to larger depth (e.g., 200 cm), is necessary in order to be able to detect compactions in deeper parts of the soils or changes in soil water content. It would be necessary for the software to recognize this automatically and to process the data accordingly. Moreover, the increased problem of rod friction should be considered
- since the soil water content influences the penetration resistance very strongly and is also very variable in terms of time and space, a simultaneous determination of the water content or the water tension over the entire measuring depth would be a major extension of the mobile electronic field penetrometer

9.3. Using a penetrometer

The following remarks assume making measurements with manual penetrometer and measuring penetration resistance vertically down from the soil surface, but are equally applicable to the use of analog or digital penetrographs. Considering the difference in scale, the vertical face of a soil profile may be studied using a mini-penetrometer in a comparable way.

9.3.1. Material required

In order to measure penetration resistance of a soil with a manual penetrometer the following material is always required:

- manual penetrometer with accessories, such as wrenches, assorted tips, replacement tips, template to identify tip wear, cleaning cloth, ecologically safe oil to wipe the rods and grease the threads
- outdoor-capable writing things, among them (paper or digital) tables, lists or sheets for writing down the data (always keep the metadata in mind): who measured, where and when was the measurement made, which equipment was used. Additionally, don't forget to make note of abnormalities of soil, weather, field, plant growth etc.
- large scale maps, perhaps a GPS receiver of sufficient accuracy to locate measuring spots directly

Where lines of spots or grids are to be measured, additionally consider:

- marker sticks; if measurements will be repeated, some material to mark the locations long term by well documented survey work, by magnets or other objects, placed below the plow depth
- tension lines, ideally with marks that indicate the intended distance of measuring spots, for soil profiles, particularly a 2 dimensional net of such tension lines

9.3.2. Prerequisites for such measurements

Penetration resistance of soils is influenced by numerous factors, hence it is mandatory to document the conditions of the actual measurement as accurately as possible.

To obtain consistent result, which are comparable among each other, all measurements of a campaign must be made with the same tip. It is recommended to check, what the smallest tip is that is able to exploit the measurement range of the penetrometer across all scheduled measurements most completely.

During a series of measurements series, no modification of the soil may occur (e.g., by plowing), because this adversely affects the measurements. Where repeat measurements are planned, it makes sense to note the soil condition when measurements are made and to review these when comparing the data. Soil water content should not vary, therefore, it is unwise to start measurements shortly after precipitation, before the distribution of water has reached a quasi-equilibrium state. There are limits imposed by the drying of the soil. For example, most soil on which sugar beets are grown are too hard in late summer to allow usable penetrometer measurements.

Naturally aggregated soils, or field soils, which have settled after being tilled, may be water saturated after rainfall, so that the aggregates or the natural textures collapse on just the slightest contact with the probe tip. Obtaining representative results from such soils (even with a large probe tip) is impossible. Ideally, penetration resistance is measured on crumbly soils with a water contents of between 80 to 100% of field capacity (FC). Such conditions are most commonly met a few days after the end of extended precipitation.

Penetration resistance data are difficult to compare, as the water content at depth influences them considerably. An absolute datum of penetration resistance is, generally speaking, less significant than the relative comparison of these. Insofar, penetration resistance is a quantity that cumulates the soil properties discussed above and its determination is always help- and useful where heterogeneity of soil parameters may be inferred from other data (loess thickness of 10 dm, high organic carbon (OC) contents down to plowing depth, no modification of the soil water content by precipitation in the preceding days, etc).

9.3.3. Planning measurements

Before starting to measure, the following questions should be addressed and answered:

- Are the measuring locations clearly marked on a map, cadastral plan or similar?
- Do weather conditions allow continuous measurement and the completion of all measurements?
- Which measuring tip will be used?
- Up to which depth are measurements to be conducted?
- Are there depth ranges requiring particular attention?
- Will the measurements be in random locations, along lines, or on grid points?
- What is the distance between measurements?
- What is the dependent variable sought?
- Maximum value per decimeter penetration
- Depth and value of maximum penetration resistance
- Average value of penetration resistance (penetrograph) for given depth ranges?
- If repeat measurements are planned, the same questions have to be answered

9.3.4. Conducting a measurement

The penetration resistance value is artificially increased if the following aspects not honored:

The (manual) penetrometer must be pushed *normally* into the soil at a *constant rate, without jerks*. The normal orientation relative to the soil surface or face is necessary to prevent rubbing (increased resistance) of the rod along the surface of the hole (which has a greater diameter, because of the thicker tip).

Within linear measurements or on a grid problems or errors due to stones, roots, burrows, irregularly pushed penetrometer etc. may be substituted by compensating measurements 10 or 15 cm away from the intended measurement location (this does not have to be documented).

Measuring locations at distances of 0.5 to 1.0 m of each other usually result in independent lines of penetration resistance with depth. They are statistically independent like exceeding the "range" distance in semi-variograms. Measuring locations less than 0.5 m result in much more similar penetration resistances. For example, the Swiss Standard Organisation postulates a distance of at least 1.0 m in their standard SN 670316a to reach independence.

Where hand penetrometers are used, it is recommended to have a second person on the field to write down the measured values and any abnormalities concerning the weather or the soil.

The remarks in this section apply equally to the use of Lightweight penetrometers for measurements from the soil surface to depth.

9.4. Data analysis

The following considerations on data analysis assume that a (manual) penetrometer was used to measure maximum resistances per 10 cm depth interval down to a depth of 1 m and the spacing of measurement points (here 20) was equal (either linear or on a grid).

To represent the data, it is recommended to tabulate them, 10 cm depth intervals horizontally, location number vertically, and report the sum of the penetration resistance values in the bottom line. Where measurements were made on a grid, it is recommended to compile such a table for every vertical *and* horizontal line of the grid, in order to be able to better identify directional distribution patterns.

Table 9.2. Representation of measured penetration resistances. 10 depth intervals (in dm) at 20 locations (A-T), at distances of 0.5 m; data in MPa * 10 (better to read). a) as measured values and their sum (bottom line); b) visualized as color-classified value; c) as cumulative value; d) as mean value; e) mean soil physical parameters per depth interval.

a) Measured values in MPa * 10 at 20 locations (A-T) with distances of 0.5 m.

Depth	Location and distance along transect (m)																			
	A	B	C	D	E	F	G	H	I	J	K	L	M	N	O	P	Q	R	S	T
	13	19	13	9	19	19	9	13	18	13	16	11	11	15	19	10	15	18	11	18
1	23	23	24	22	22	24	21	18	22	20	19	18	22	20	19	17	23	20	20	21
2	22	21	20	19	19	19	21	18	19	18	22	16	23	15	20	18	24	20	25	19
3	19	20	19	17	13	19	16	22	16	20	17	13	19	20	17	15	22	15	25	21
4	20	22	17	16	16	19	14	18	17	20	16	14	19	14	19	13	24	16	18	24
5	21	24	17	14	15	21	12	17	20	22	19	17	20	19	18	16	20	18	19	27
6	30	25	24	22	23	24	24	17	22	27	22	28	25	28	18	28	16	25	23	32
7	34	38	36	30	27	38	36	29	31	33	33	35	29	31	32	33	29	30	33	34
8	99	39	39	38	37	99	38	36	36	36	39	39	35	39	34	36	39	33	39	99
9	99	99	99	99	39	99	99	38	38	39	99	99	39	99	33	99	99	37	99	99
10	380	330	308	286	230	381	290	226	239	248	302	290	242	300	229	285	311	232	312	394
Sum	380	330	308	286	230	381	290	226	239	248	302	290	242	300	229	285	311	232	312	394

Soil mechanical parameters

b) Colored classification of measured values from a).

0 to 1.0 MPa, given in MPa * 10, i.g.: 0 to 10	
1.1 to 2.0	11 to 20
2.1 to 3.0	21 to 30
3.1 to 4.0	31 to 40
<= 4,1	>= 41

Location and distance along transect (m)

Depth	A 0.5	B 1.0	C 1.5	D 2.0	E 2.5	F 3.0	G 3.5	H 4.0	I 4.5	J 5.0	K 5.5	L 6.0	M 6.5	N 7.0	O 7.5	P 8.0	Q 8.5	R 9.0	S 9.5	T 10.0
1	13	19	13	9	19	19	9	13	18	13	16	11	11	15	19	10	15	18	11	18
2	23	23	24	22	22	24	21	18	22	20	19	18	22	20	19	17	23	20	20	21
3	22	21	20	19	19	19	21	18	19	18	22	16	23	15	20	18	24	20	25	19
4	19	20	19	17	13	19	16	22	16	20	17	13	19	20	17	15	22	15	25	21
5	20	22	17	16	16	19	14	18	17	20	16	14	19	14	19	13	24	16	18	24
6	21	24	17	14	15	21	12	17	20	22	19	17	20	19	18	16	20	18	19	27
7	30	25	24	22	23	24	24	17	22	27	22	28	25	28	18	28	16	25	23	32
8	34	38	36	30	27	38	36	29	31	33	33	35	29	31	32	33	29	30	33	34
9	99	39	39	38	37	99	38	36	36	36	39	39	35	39	34	36	39	33	39	99
10	99	99	99	99	39	99	99	38	38	39	99	99	39	99	33	99	99	37	99	99
Sum	380	330	308	286	230	381	290	226	239	248	302	290	242	300	229	285	311	232	312	394

9. Penetration resistance

c) Values from a) cumulated to depth

Depth												Location and distance along transect (m)								
	A	B	C	D	E	F	G	H	I	J	K	L	M	N	O	P	Q	R	S	T
	0.5	1.0	1.5	2.0	2.5	3.0	3.5	4.0	4.5	5.0	5.5	6.0	6.5	7.0	7.5	8.0	8.5	9.0	9.5	10.0
1	13	19	13	9	19	19	9	13	18	13	16	11	11	15	19	10	15	18	11	18
2	36	42	37	31	41	43	30	31	40	33	35	29	33	35	38	27	38	38	31	39
3	58	63	57	50	60	62	51	49	59	51	57	45	56	50	58	45	62	58	56	58
4	77	83	76	67	73	81	67	71	75	71	74	58	75	70	75	60	84	73	81	79
5	97	105	93	83	89	100	81	89	92	91	90	72	94	84	94	73	108	89	99	103
6	118	129	110	97	104	121	93	106	112	113	109	89	114	103	112	89	128	107	118	130
7	148	154	134	119	127	145	117	123	134	140	131	117	139	131	130	117	144	132	141	162
8	182	192	170	149	154	183	153	152	165	173	164	152	168	162	162	150	173	162	174	196
9	281	231	209	187	191	282	191	188	201	209	203	191	203	201	196	186	212	195	213	295
10	380	330	308	286	230	381	290	226	239	248	302	290	242	300	229	285	311	232	312	394

d) Average values to depth from a)

Depth												Location and distance along transect (m)								
	A	B	C	D	E	F	G	H	I	J	K	L	M	N	O	P	Q	R	S	T
	0.5	1.0	1.5	2.0	2.5	3.0	3.5	4.0	4.5	5.0	5.5	6.0	6.5	7.0	7.5	8.0	8.5	9.0	9.5	10.0
1	13	19	13	9	19	19	9	13	18	13	16	11	11	15	19	10	15	18	11	18
2	18	21	19	16	21	22	15	16	20	17	18	15	17	18	19	14	19	19	16	20
3	19	21	19	17	20	21	17	16	20	17	19	15	19	17	19	15	21	19	19	19
4	19	21	19	17	18	20	17	18	19	18	19	15	19	18	19	15	21	18	20	20
5	19	21	19	17	18	20	16	18	18	18	18	14	19	17	19	15	22	18	20	21
6	20	22	18	16	17	20	16	18	19	19	18	15	19	17	19	15	21	18	20	22
7	21	22	19	17	18	21	17	18	19	20	19	17	20	19	19	17	21	19	20	23
8	23	24	21	19	19	23	19	19	21	22	21	19	21	20	20	19	22	20	22	25
9	31	26	23	21	21	31	21	21	22	23	23	21	23	22	22	21	24	22	24	33
10	38	33	31	29	23	38	29	23	24	25	30	29	24	30	23	29	31	23	31	39

e) Average soil physical parameters to respective depth.

Depth	BD (g cm^{-3})	PV (vol. %)	Θ (vol. %)	Soil type[#]
1	1.44	45.8	4.3	Su2
2	1.44	45.6	3.5	Su2
3	1.49	43.7	10.3	Su2
4	1.27	52.2	30.0	Su2
5	1.45	45.1	6.5	Su2
6	1.23	53.7	15.2	Su2
7	1.56	41.4	6.2	Su4
8	1.66	37.5	12.0	Su4
9	1.74	34.2	11.4	S
10	1.62	38.8	11.6	S

BD = bulk density; PV = total pore volume; Θ = actual water content;
[#] according to mapping manual or DIN 4220

The headline of each table lists the depth (dm) in the first column and the locations in the subsequent columns, possibly with distance from the starting point. Per location the maximum value in the depth interval is shown by **bolding** or framing the corresponding table cell and the sum of the values per location is indicated in the bottom line. Furthermore, table cells are colored or greyed, according to which range of resistances the measured value falls into. Commonly, five ranges of values are distinguished, often corresponding to the usual range of penetrometers of 0–5 MPa in 1 MPa steps (Table 9.2b). Depending on the data, other subdivisions may of course be more useful. The same values may be displayed as isolines of penetration resistance with the aid of suitable computer programs.

Note, that non-equidistant spacing should be indicated in the table and visible on the isoline plot. Furthermore, a key listing the colors and the range of values they correspond to should be provided, with the hint, that, given quadratic table cells, this corresponds to a vertical exaggeration of 1:10.

By arranging the table as outlined above and bolding maximum values, the first steps of further data analysis have already been made. Further approaches are:

- computing sums for other depth ranges (1–3, 4–5, 5–8 dm) etc.
- mean values may be computed from the sums and the values in these depth ranges may then be shown as deviation from the mean value.
- a table which cumulates values down to the depth of interest, so that the line at 10 dm corresponds to the bottom line which contains the summed up values.
- a table which shows the measured values as percentages of the cumulated values in the bottom line. The maximum depth line then corresponds, just as the bottom line to 100%. This table can also be color coded or represented as an isoline graph.

In most cases, the interpretation of penetration resistance data has to be done qualitatively. Therefore, it is recommended to compare preferably measurements conducted under comparable conditions, and to compare relative values, rather than absolute ones. For some cases it is additionally possible to compare absolute measurements with limits and to derive properties and usability of the soil.

Using the suggested methods to represent these data, it is possible to portray the homogeneity or heterogeneity of depth distribution of penetration resistance in detail.

Some depth profiles of penetration resistance may reflect the effect of one known factor on soil. They may also be used to derive statements about the accuracy of such measurements in terms of variance.

Figure 9.5a shows the effect of repeated crossings of a soil by heavy machinery on the distribution of penetration resistance with depth. On the one hand, in 2 dm depth penetration resistance increases with number of crossings. On the other hand, maximum of penetration resistance goes deeper following eight crossings. Additionally, increasing compaction can be observed below 5 dm depth.

Figure 9.5b shows one depth profile of penetration resistance with 1 cm resolution and additionally its standard deviation (1* sigma) Obviously, in two to four decimeter there is an area of similar penetration resistance and standard deviation, in about eight decimeter followed by an area with increasing resistance and higher standard deviation, below it decreasing values and deviations.

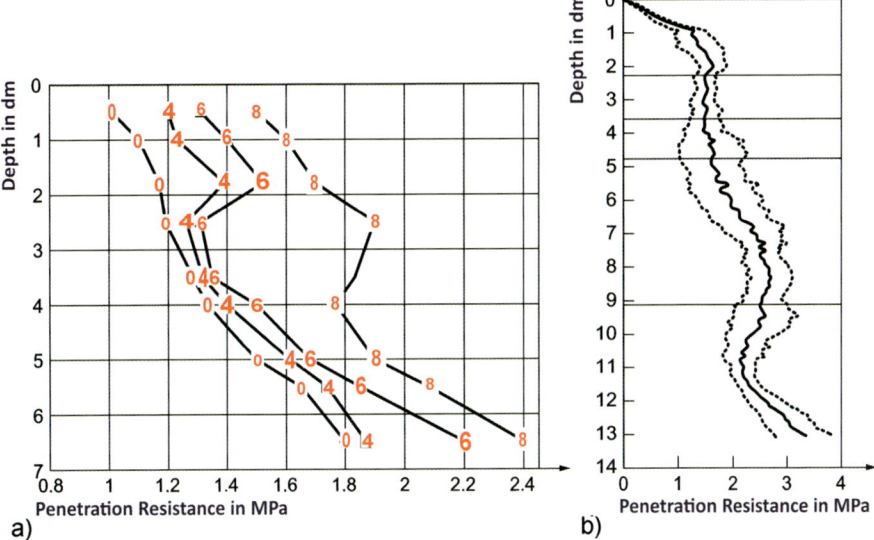

Fig. 9.5. Depth profiles of penetration resistance in soils. a) comparing the effect of frequency of crossings: 0, 4, 6 and 8 times; b) showing high resolution mean values of penetration resistance and their associated standard deviation (1* sigma).

The key factors listed in section 9.1.1, which control penetration resistance may be assigned to three groups:
- temporal variation of soil water contents (soil water tension)
- grain size distribution and fraction of coarse fragments (gravel, cobble, or boulder)
- bulk density and possible soil compaction and cementation

Where large scale soil maps or local soil mapping data are available, they may provide hints as to the reason of the spatial distribution of penetration resistance, in terms of the three groups of factors above. Where no maps are available, exemplary soil auger samples taken close to measuring locations with distinct depth profiles may be surrogates in order to interpret the measured distributions.

Obtaining reliable information on the rootability of a soil for crop plants is as difficult, as it is to use this data to assess the rootability of trees in urban settings or parks. The reason is, that a host of other factors influencing rootability and that roots do not just grow in tip direction but also radially, which a penetrometer is not designed to measure.

Nevertheless, obtaining a penetration resistance of 2.0 MPa and greater at a soil water content of 80 to100% of field capacity (FC) may be an indicator of soil compaction which corresponds to root growth rates at only 50% of optimal conditions. Values of 3 MPa and above will most certainly result in detrimental effects on plant growth and correspondingly reduced crop yields, or even complete crop loss.

As penetration resistance of soils is controlled by so many factors, its integral measurement with a penetrometer is the method of choice to quantify areal and temporal (soil water content) homogeneity of penetration resistance as well as to select representative areas for examinations.

All in all, despite the numerous factors influencing the penetration resistance, the hand penetrometer affords a cost effective and quick measurement of this soil parameter, the data of which are easily analyzed and represented.

9.5. References

AFNOR, 2000. AFNOR Standard XP P94-105: Soils: investigation and testing – Inspection of compaction quality – Method using a variable energy dynamic penetrometer – Principle and method for calibrating the penetrometer – Exploitation of results – Interpretation (In France with English abstract), ISSN 0335-3931.

ASTM 1999. ASTM Standard D1558: Standard Test Method for Moisture Content Penetration Resistance Relationships of Fine-Grained Soils.

ASTM 2005. ASTM Standard D3441: Standard Test Method for Mechanical Cone Penetration Tests of Soil.

ASTM 2010. ASTM Standard D4633: Standard Test Method for Energy Measurement for Dynamic Penetrometers.

DIN 2011. DIN Standard DIN 19662: Soil quality – Field tests – Determination of soil penetration resistance by means of a hand held penetrometer. (In German with English abstract.) – Beuth, Berlin.

SNV 1994. SWISS Standard SN 670316a-1994-06: Investigations of soils – CRB-Penetrometer, field test. (In German, France with English abstract).

Hartge, K.H., Bohne, H., Schrey, H.P., Extra, H., 1985. Penetrometer measurements for screening soil physical variability. Soil and Tillage Research 5: 343–350.

Hartge, K.H., Horn, R., 2016. Essential Soil Physics. An introduction to soil processes, functions, structure and mechanics. editors: R. Horton, R. Horn, J. Bachmann and S. Peth, Schweizerbart, Stuttgart.

Schrey, H.P., 1987. A qualitative and quantitative representation of penetration resistance distributed over depth. Mitteilungen Deutsche Bodenkundliche Gesellschaft 55: 239–244 (In German).

Index

accuracy 12
activity 48
actual acidity 53
air pressure potential 84
algae growth 115
alkali ion activity 51
aluminum saturation 53
ammonium 125
antimony (Sb) electrode 45
ASTM D3385-09 140
atmosphere 13

bacteria 126, 127
bentonite 16
borehole-method 134, 151, 153
boundaries 13
boundary layer 171
breakthrough curve 110, 113
budget 9
bulk density 15, 191
bulk electrical conductivity 71
bulldozer-effect 186
bypass flow 114

cable 15
cable tester 72
calibration plate 169
calibration vessel 183
capacitance device 71
capillary wick 99, 116
characteristic length 10
chromium 126
colloid 105, 126, 127
color tracer 183
colorimetry 45
complex refractive index model 75
conservative tracer 115
constant head method 137
constant head permeameter 157
cosmic-ray neutron probe 77
cosmic-ray particles 77
cost 9
cumulative infiltration 135

data logger 25, 56
deep percolation 133
deep pressure vacuum lysimeter 99
delta pH 55
depth profile 204
dielectric medium 71
DIN 19682-7 140

DIN 19682 134
DIN 19682-8 152
DIN 4094-3 193
dirty electrode 51
dissolved load 171
dissolved organic carbon (DOC) 105, 126, 127
dissolved organic matter (DOM) 105, 126, 127
double-box slope-infiltrometer 147
double-ring infiltrometer 139

Edelman auger 151
effective stream power 185
electrical impedance (EI) sensor 72
electrical resistivity tomography 77
electrode contamination 34
electrode potential 21
electromagnetic induction 77
electromagnetic soil moisture sensor 69
electron acceptor 19
electron donor 19
electrons 19
equilibrium condition 34
equilibrium tension plate lysimeter (ETPL) 108
eroded material 167
erosion by water 165
erosion research 165
erosion risk classification scheme 168
evapotranspiration 13
exchange acidity 53
experimental setup 14, 167
experimental soil erosion research 166
extraction tube 100

falling head method 137
field capacity 109
flooded infiltration 134
flow velocity of water 183
free drainage lysimeter 121
frequency domain (FD) sensors 72

glass electrode 47
glass measuring electrode 56
GPS-receiver 197
grain size 191
gravimetric soil water content 69
gravitational potential 83
grid sampling with bulking 61
grid spacing 61
ground-penetrating radar 77
groundwater 133
groundwater level 151

Guelph-permeameter 134, 148, 149, 150, 151
gully erosion 165

hardware damage 51
heavy metals 126
heterogeneity 10, 204
homogeneity 204
honeycomb 171
hood infiltrometer 142
horizon 14
hose-like duct (transformer) 171
hydraulic conductivity 142, 152
hydraulic potential 85
hydrogeophysical method 69
hydrological shortcut 15
hydronium ion activity 43
hydrophobicity 133
hydrostatic pressure potential 84

impedance 72
infiltration 133
infiltration capacity 135, 142
infiltration class 147
infiltration rate 135, 142
initial rate of infiltration 135
inner electrolyte solution 47
in-situ experiment 167
installation 13
installation depth 14
instrumentation 10
interference from alkali ions 51
ion sensitive field effect transistor (ISFET) 45
irrigation tunnel 173
ISO 22476-2 193

Kopecky-ring 70
Künzelstab 193

lab experiment 167
large scale-method 134
laser distance meter 179
leachate collection efficiency 119
leached mass fraction (LMF) 108
lightweight penetrometer 193
load cell 121, 123
long-term effect 45
LoRa (long range) 75
lysimeter 99

macropore 142
mapping strategy 61
mass of eroded material 185
matric potential 84
mean concentration 17
mean value 17

measurement plate 179
measuring bridge 179
measurement location, measuring location 9, 179
membrane 100
memory effect 61
microorganism 126, 127
mid-infrared light spectroscopy (MIRS) 52
mobile electronic field penetrometer 198
monitoring plot 62
monitoring station 9
multicompartment sampler 108

Nernst equation 21, 49
nitrate 125
non-invasive measurement techniques 77
nozzle 169
nylon 105, 126

offset voltage 50
Ohm's law 29
oil-to-solution ratio 52
on-the-go soil pH sensor 52
optode 46
organic pollutant 105, 126
organic rich soil 75
oscillating circuit 72
osmotic potential 84
overburden potential 85
oxidation 19
oxidation pH 55

pan lysimeter 99
passive wick fluxmeter 119
PE 105, 126
pe value 21
penetration resistance 191
penetrograph 196
penetrologger 196
penetrometer 192
percussion penetrometer 194
permeability rate 148
pesticide 105, 126
petrophysical relationship 71
pF-value 84
pH 22
pH meter 56
pH sensitive silicate glass membrane 47
pH value 44
pH_{CaCl_2} 54
pH_{H_2O} 54
pH_{KCl} 54 55
pH_{NaF} 55
phosphate 125
pH-sensitive fluorophores 46
pit 15

Index

plant available water 83
platinum (Pt) electrode 20
plot 169
plunge pool dynamic 186
polychlorinated biphenyls (PCBs) 105, 126
polycyclic aromatic hydrocarbons (PAHs) 105, 126
polytetrafluoroethylene 125
porous candle 99
porous ceramic 100
porous cup 99
porous plate 99
porous tube 99
portable wind and rainfall simulator (PWRS) 167
post-processing 17
potential 83ff
potential (Φ) 148
potential acidity 53
precipitation 13
precision 12
preferential flow 100, 115
PTFE 125
pulse generator 72
Pürckhauer auger 151
PWRS field experiment 173

reactive substance 122
redox buffer solution 27
redox potential 19
reduction 19
reference electrode 56
reference value 12
relative dielectric permittivity 71
remote sensing 75
replicate 10
representative area 206
REV (representative elementary volume) 10 f
rill erosion 165
rill experiment 168, 175
ross-contamination 115
rubber mat 177
runoff 167, 171
runoff measuring station 178

salt bridge 25, 56
sample bottle 101
saturated hydraulic conductivity 133, 148
saturated soil 133
saturated water conductivity 157
saturation 114
scale 10
sediment load 171
sediment load concentration 167
sediment trap 172
seepage face 109

seepage flow 114
sensor network 72, 73
shaft 15
shear stress 185
short- to mid-term variation 44
short-term field measurement 52
simulation of the effects of rain on a dry soil surface 174
simulation of the effects of rain on a moist soil surface 174
simulation of the effects of simultaneous wind and rain 174
simulation of wind erosion 174
single-ring infiltrometer 135
single-ring infiltrometry 135
sintered glass 100
sintered metal 100
small portable rainfall simulator 167
small-scale method 134
SN 670316a 200
sodium error 51
soil character 44
soil column 119
soil compaction 206
soil ecosystem service 43
soil erosion 165
soil moisture 69
soil qualifier 44
soil temperature 11
soil water content 69, 191
soil water content sensor 69
solidification 191
sources of errors and abrasion 50
spatial heterogeneity 9, 101
spatial resolution 13
spatially dense sampling 62
splash erosion 165
stability criterion 53
standard deviation 17
standard electrode potential 21
standard hydrogen electrode 21
standardization of rainfall simulation setups and experiments 171
steepness 49
stream power 185
subsurface flow 133
suction cup 99
suction cup activity domain (SCAD) 107
suction cup extraction domain (SCED) 107
suction cup sampling area (SCSA) 107
suction lysimeter 99, 121
suction probe 99
sulfidic material 55
sulfuric horizon 55
surface free energy 84

surface runoff 124, 133
suspension effect 45, 50

tape measure 177
targeted sampling with bulking 62
Teflon® 125
temperature compensation factor 50
temperature effect 50, 73
tensiometer 86ff
tensiometer potential 86
tension infiltration 142
tension infiltrometer 142, 145
tension infiltrometry 142
time domain reflectometry (TDR) 71
time domain transmission (TDT) 71, 73
tin oxide (SnO_2) transparent electrodes 45
topographic profile 179
transport capacity of the rill 184
transport rate 185
trench 15
trueness 12
tube 15
two-blade axial propeller 171
two-probe glass electrode 49

ultrasonic sensor 179
unit length shear force 185
unit stream power 185
unsaturated hydraulic conductivity 142, 145

unsaturated soil 133

vacuum 101
vacuum extractor 99
vanadium 126
variation 11
virus 126, 127
visible and near-infrared light spectroscopy (VNIRS) 52
volatilization 120
voltmeter 25
volumetric soil water content 69

water conductivity 133
water level 179
water potential 86
water saturation 133
waveforms 77
wick lysimeter 116
wilting point 83
wind effect 124

xenobiotic 105, 126

ZigBee 73

Quantifying the Environment

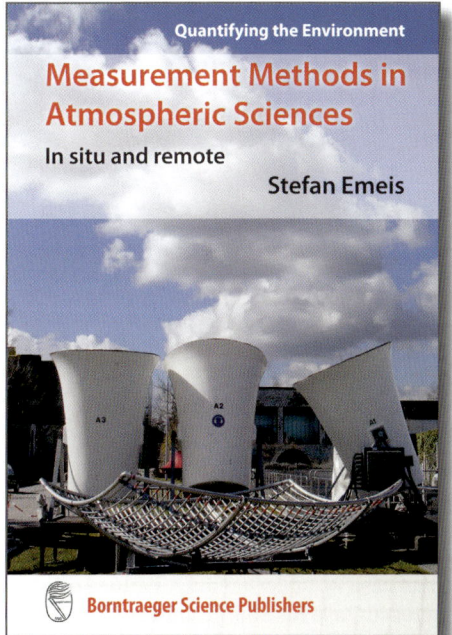

Quantifying the Environment

Stefan Emeis

Measurement Methods in Atmospheric Sciences

In situ and remote

2010. XIV, 257 pp., 103 colour figs, 28 tables
17 x 24 cm, hardcover
ISBN 978-3-443-01066-9   68.– €

Measurement Methods in Atmospheric Sciences provides a comprehensive overview of in-situ and remote sensing measurement techniques for probing the Earth's atmosphere. The methods presented in this book span the entire range from classical meteorology via atmospheric chemistry and micro-meteorological flux determination to Earth observation from space. Standard instruments for meteorological and air quality monitoring methods, as well as specialized instrumentation predominantly used in scientific experiments, are covered. The presented techniques run from simple mechanical sensors to highly sophisticated electronic devices. Special emphasis is placed on the rapidly evolving field of remote sensing techniques. Here, active ground-based remote sensing techniques such as SODAR and LIDAR find a detailed coverage. The book conveys the basic principles of the various observational and monitoring methods, enabling the user to identify the most appropriate method.

The book is of interest to undergraduate and graduate students in meteorology, physical geography, ecology, environmental sciences and related disciplines as well as to scientists in the process of planning atmospheric measurements in field campaigns or working on data already acquired. Practitioners in environmental agencies and similar institutions will benefit from instrument descriptions and the extended lists in the appendix.

Borntraeger Science Publishers
Stuttgart

Johannesstr. 3a, 70176 Stuttgart, Germany. Tel. +49 (711) 351456-0 Fax. +49 (711) 351456-99
order@borntraeger-cramer.de   www.borntraeger-cramer.de

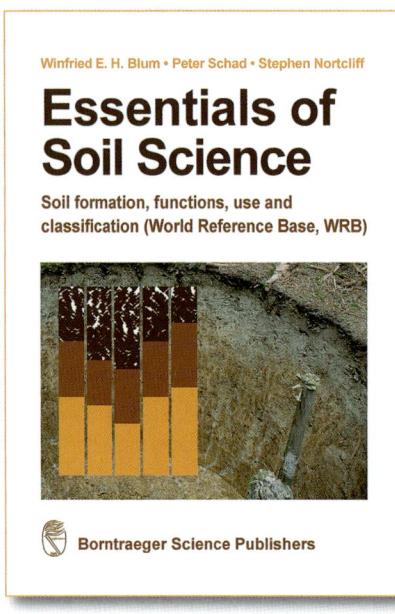

Winfried E. Blum • Peter Schad • Stephen Nortcliff

Essentials of Soil Science

Soil formation, functions, use and classification (World Reference Base, WRB)

2018. 171 pp., 101 figures, 22 tables, 17 x 24 cm, softcover

ISBN 978-3-443-01090-4   27.90 €

borntraeger-cramer.com/9783443010904

This book is an introduction to soil science and describes the development of soils, their characteristics and their material composition as well as their functions in terrestrial and aquatic environments. Soil functions include the delivery of goods and services for the human society, such as food, clean water, and the maintenance of biodiversity. The book is profusely illustrated with many coloured figures and tables to accompany the text and ease its understanding. Particularly, the chapter on soil classification, based on the World Reference Base for Soil Resources (WRB), includes numerous coloured pictures to facilitate understanding the characteristics of particular soil types. Chapters on soil protection and remediation as well as on soil monitoring and the history of soil sciences conclude the book together with a very comprehensive alphabetical index, allowing for a quick and easy orientation about the most important terms in soil sciences.

This very concise and at the same time comprehensive publication addresses all those, who want to orient themselves about soils, their functions, their importance in terrestrial and aquatic environments and their contribution to the actual and future development of the human society, such as teachers, practitioners and students in the fields of agriculture, forestry, gardening, terrestrial and aquatic ecology and environmental engineering, and of course, beginning students of soil science.

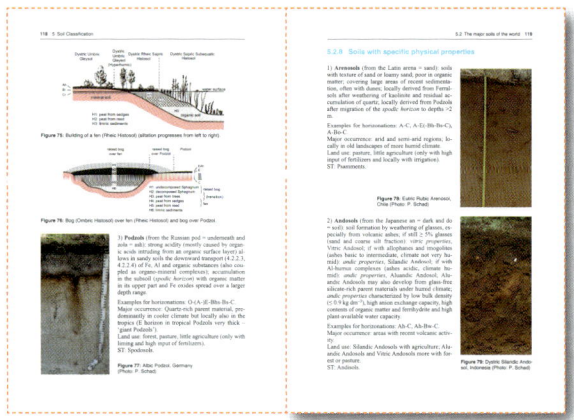

Borntraeger Science Publishers

Johannesstr. 3A, 70176 Stuttgart; Germany
Tel. +49 (711) 351456-0   Fax. +49 (711) 351456-99
order@schweizerbart.de   www.borntraeger-cramer.com